William Penn
Process Realism in Physics

Process Thought

Edited by
Nicholas Rescher, Johanna Seibt, Michel Weber

Volume 28

William Penn

Process Realism in Physics

How Experiment and History Necessitate a Process Ontology

DE GRUYTER

ISBN 978-3-11-221412-1
e-ISBN (PDF) 978-3-11-078251-6
e-ISBN (EPUB) 978-3-11-078267-7
ISSN 2198-2287

Library of Congress Control Number: 2023933436

Bibliographic information published by the Deutsche Nationalbibliothek
The Deutsche Nationalbibliothek lists this publication in the Deutsche Nationalbibliografie;
detailed bibliographic data are available on the internet at http://dnb.dnb.de.

© 2025 Walter de Gruyter GmbH, Berlin/Boston
This volume is text- and page-identical with the hardback published in 2023.
Printing and binding: CPI books GmbH, Leck

www.degruyter.com

Acknowledgements

This work is the result of many fruitful conversations and exchanges with members of the philosophy community, and could not have been completed without every single one. A few are worth special mention. John Norton was key in helping me to develop both my writing and the specifics of the ideas contained in this book. Johanna Seibt greatly improved my understanding of process ontology, and was integral in my ability to apply it to scientific theory and practice. Jim Woodward was integral throughout my career in pushing back against my ideas, and helping them to develop into more nuanced arguments. Erica Shumener was the central figure in my development as a metaphysician. Debra Nails helped me realize for the first time the difficulties of substance ontologies. In addition, conversations with Colin Allen, Nuhu Osman Attah, Gal Ben-Porath, Agnes Bolinska, Nora Mills Boyd, Anjan Chakravartty, Hasok Chang, Mazviita Chirimuuta, Kathleen Creel, Siska DeBaerdemaeker, Marina DiMarco, John Earman, Marian Gilton, Sara Green, Mahi Hardalupas, Jenan Ismael, Michel Janssen, Shahin Kaveh, Kareem Khalifa, Eleanor Knox, James Lennox, Michela Massimi, Dana Matthiessen, Sandra Mitchell, Nedah Nemati, Rose Novick, Cailin O'Connor, Paolo Palmieri, Morgan Thomson, David Wallace, Ken Waters, James Weatherall, Porter Williams, and Jennifer Whyte were essential for developing this work. Lastly, I thank my family for all their support.

Preface

It is commonly assumed in philosophy that reality is composed of things, the holders of definite properties, the bearers of definitions, and the grounds for certain, absolute truths. We take for granted that science and its history has been aimed at developing our understanding of these absolute and static things. Great advances in the history of science are reformulated to fit this image of discovering the true nature of the world. From genes to particles to spacetime structure to molecular makeup, we treat science as if it will eventually tell us how the world is, absolutely, eternally, and definitely.

However, the history of science, and its foundation in empirical practice tell a different story. Rather than definite properties of the world, science reveals dynamic and contextual features. Rather than absolute truths, science tells of evolving understanding and modeling practice. Rather than static structural relations, science outlines a world of interactions. Fundamentally, because science is not primarily a description of the world, but rather a method for interacting with it and learning from it, science allows us to know not a world of things, but a world of processes.

Drawing on both conceptual analysis of the empirical practice of science, particularly physics, and the history of this practice, this book aims to articulate exactly what the dynamic world of science looks like. It outlines two elements of this picture. First, an ontology of science rooted in the nature of experiment. Second, an epistemology of the inferences that scientists and philosophers alike are allowed to make in the process of experimenting on and modeling the physical world. The book argues that this new way of looking at the world—the process-realist way—resolves longstanding philosophical problems and directs new research in science. Finally, the book presents process realism as another iteration on a worldwide tradition of process philosophy finding roots in a diverse array of philosophies and scientific communities alike. Thus, the book presents process realism as the culmination of the philosophies of the world and the international community of practicing scientists. Process realism is what is really going on.

Contents

Acknowledgements —— V

Preface —— VII

List of Tables —— XIII

List of Figures —— XV

Introduction —— 1

Chapter 1 Continuity and Process Realism —— 3
1.1 Introduction —— 3
1.2 Defining "Process" for Use in the Philosophy of Science —— 5
1.2.1 Processes vs. Things —— 6
1.2.2 What Is a Process? —— 8
1.3 A Thought Experiment: First-Level Process-Realist Commitment —— 12
1.3.1 The Unchanging Room —— 12
1.3.2 Without Process, No Observation —— 15
1.4 The Continuity Argument —— 17
1.4.1 The Simple Continuity Argument —— 18
1.4.2 The Complex Continuity Argument —— 21
1.5 Conclusion —— 31

Chapter 2 Processes Underlie Processes —— 34
2.1 Introduction —— 34
2.2 Underlier Arguments and Their Types —— 36
2.2.1 Underliers of Experimental Practice: Stability within and between Experiments —— 37
2.2.2 Stability of Descriptions and Models —— 58
2.2.3 Manifest and Assumed Stability —— 72
2.3 General Refutation: From Negation to Position —— 84
2.3.1 The Negative: Refuting Underlier Arguments Algorithmically —— 84
2.3.2 The Positive: Explaining Stability —— 86
2.4 Conclusion —— 88

Interlude: Two Shifts in Method —— 91

Chapter 3 The Candle Flame: A Process-Realist Analysis —— 93
3.1   Introduction —— 93
3.2   The Flame and Phlogiston, the Thing Realist's Posit —— 95
3.3   Explaining Things Away —— 99
3.3.1 What Do We See in the Flames? —— 99
3.3.2 What Do We Explain about the Flames? —— 102
3.3.3 Explaining What Needs Explaining —— 103
3.4   What Explanatory Role, Things? —— 115
3.4.1 New Thing Terms Appear in Our Explanations ... —— 115
3.4.2 ... But Things Play No Role in Our Explanations —— 116
3.5   Conclusion —— 122

Chapter 4 Perrin's Argument: A Robustness Argument for Processes, Not Things —— 126
4.1   Introduction —— 126
4.2   Perrin's Intuition Pump: The Bath and Cascading Fluid Motion —— 127
4.3   Perrin's Historical/Eliminative Argument —— 129
4.4   Perrin's Precise Arguments —— 132
4.4.1 Argument 1: Qualitative Robustness —— 132
4.4.2 Argument 2: Quantitative Robustness —— 135
4.4.3 Summary —— 144
4.5   Conclusion —— 145

Chapter 5 Models of the Nucleus: Incompatible Things, Compatible Processes —— 147
5.1   Introduction —— 147
5.2   Many Models, Divergent Things —— 149
5.2.1 The Liquid-Drop Model —— 150
5.2.2 The Shell Model —— 153
5.2.3 Incompatibility —— 155
5.3   The Features of the Nucleus Are Processes —— 157
5.3.1 Formal Features of the Nucleus: Balanced Dynamics —— 159
5.3.2 Material Features of the Nucleus —— 166
5.3.3 Reestablishing Compatibility —— 170
5.4   Conclusion and Prospectus —— 171

Summary and Prospectus —— 174

Bibliography —— 175

Index —— 194

List of Tables

Table 1: A list of features of the candle flame that Faraday seeks to explain, grouped into three categories (Chapter 3)
Table 2: A comparison of properties and thing-claims made by the liquid-drop and shell models (Chapter 5)
Table 3: A list of formal, material, and productive features of the nucleus to be explained (Chapter 5)
Table 4: A collection of the processes that appear in both models. These collections are identical, but are used in different ways to explain different ends. (Chapter 5)

List of Figures

Figure 1: The nearest-neighbor interactions described by the liquid-drop model. Pairing and asymmetry binding forces are empirically, not theoretically motivated. (Chapter 5)

Figure 2: A depiction of the relative effects of each interaction type on the accuracy of the predicted binding energy curve. The energy of each iteration type is shown as the difference between curves. (Chapter 5)

Figure 3: Shell quantization in the shell model, showing energy-momentum shells and spin-orbit subshells. Shell closure occurs at the indicated magic numbers of nucleons: 2, 8, 20, 28, 50, 82, 126. (Chapter 5)

Figure 4: A representation of nuclear density and radius as related to nucleon number. Large differences in both properties are observed between A and C and B and C. (Chapter 5)

Figure 5: A schematic of the identifiable, token processes involved in a token scattering event as described by the liquid-drop model (Chapter 5)

Introduction

This work is dedicated to defending a novel form of scientific realism that I call pure process realism. This realism commits us to those aspects of our models and theories that represent subjectless, non-particular, dynamically contextual activities like motions, excitations, decays, fluctuations, and more. While the general ontology of subjectless, non-particular, dynamically contextual activities is the work of others, the justification I offer for this ontology in the context of scientific models, experiments, and theories is novel. In total, there are five positive arguments for process realism, intermingled with three arguments against non-process or mixed-process forms of realism involving determinate, atemporal, independent, particular things like structures, substances, particles, and (purely spatial) states.

Chapter 1 is designed to show that, so long as one commits to the reality of an observable world, one must commit to the reality of processes. This is because our mere ability to observe is predicated on the existence of processes to enable and ground these observations. Moreover, because of the details of how we observe and experiment, we can commit not merely to vague and undefined processes, but also to specific and highly specified processes contained in our scientific models. In brief, parts of our experiments will inherit processual features from the intervention, preparation, and observation activities (processes themselves) that we perform.

Chapter 2 represents the negative of Chapter 1. While we are allowed to infer the existence of processes on the basis of experiment alone, we are not allowed to commit to things. Supplementing standard antirealist arguments against the theoretical thing-posits of historical physics—like phlogiston, the four elements, mysterium, and Rutherford atoms—I also argue that any inference to things on the basis of observation is dubious. In fact, arguments for things rest on first inferring that there are processes, and then further inferring that these processes must have subjects, vehicles, or continuants (depending on the terminology of the relevant literature). These arguments that I call "underlier arguments" have existed since Aristotle, they have taken many forms, and all fail for the same reason: they presuppose that stability entails staticity. This premise is false insofar as it is not conceptually necessary, since processes can be considered stable but are not static. In fact, recognizing that stability is a relative term that depends on comparisons of processual entities allows us to subvert and co-opt underlier arguments in favor of process realism.

Chapter 3 is the first of three chapters in which I analyze the practice and history of scientific inquiry. In it, I present an extended example of the candle flame as described in both historical and contemporary chemistry and physics. I show that nowhere in these treatments are we meant to explain aspects of the system in terms of things, instead presenting a full deconstruction of thing terms into processes alone. This chapter, then, is meant to illustrate the explanatory defeat of things through the particulars of a generalizable example. The example suggests that all we need for our explanations in physics and chemistry are processes.

Chapter 4 takes this one step further. I consider the famous argument from Perrin that atoms are real, oft taken in the literature as a triumph of thing realism in the history of physics. I show that this is not accurate: a closer reading of Perrin shows that we are meant to reject realist claims about things (qua determinate, atemporal, independent, particular entities), even in this supposed argument for atoms. Instead, I reinterpret Perrin as offering a robustness argument for processes, or rather, for a specific quantifiable aspect of thermal and statistical processes like Brownian motion. Namely, I argue that Perrin is presenting evidence that thermal and statistical processes, especially processes of dispersion, can be quantified such that they achieve equilibrium when these dispersion processes all achieve the same characteristic energy.

Chapter 5 is where I present the example of nuclear models, incompatible on thing interpretations, and show that they are compatible in pure process interpretations. This argument acts as evidence that thing realism is not only historically inadequate, it is also philosophically problematic in contemporary physics. Process realism, in contrast, provides good accounts of both the history and practice of science and solves some of the problems produced by the uncritical commitment to thing ontologies.

Chapter 1
Continuity and Process Realism

1.1 Introduction

A central debate in the philosophy of science is the question of how to interpret our scientific theories and models. Scientific realism treats our theories and models as if they realistically represent the world, in one manner or another. Antirealists, in contrast, suggest that our theories and models fail to realistically represent the world, but are still epistemically fecund since they are empirically adequate. Many attempts to justify scientific realism look to the features and relations of our models and theories in order to find realist posits and claims.[1] I propose something else: since we know that our models must be empirically adequate, i.e., pragmatically describe and/or enable some class of physical experiments, we should instead look to the features of experiments to see how realist claims can be justified. If there is some form of realism that can produce realist claims as a result of taking seriously that models essentially describe experiments, then we will have produced a realism that is more robust against the standard arguments of the antirealist.

This chapter centers on a single argument at three levels of complexity for what I call pure process realism. I argue that processes—such as motions, interactions, fluctuations, etc. which I understand as non-particular, subjectless activities[2]—are necessary parts of any and all scientific ontologies. This is because observations, and therefore experiments, are physically impossible without the existence of real processes, as I will show. In every experiment, something happens or occurs,[3] and we respond to it dynamically. In this manner, I argue that the plethora of new work in philosophy of science to justify processual or process-adjacent in-

---

[1] General arguments to this effect are found in, e.g., Chakravartty (2007), Psillos (1999), Smart (1963) (these argue that models and theories give knowledge of features of the world), or in van Fraassen (1980) (who argues that models aim to give true descriptions of the world). Specific arguments for various types of realism will be discussed later, where relevant.
[2] The specifics are discussed in section 1.2. I draw my applied ontology primarily from Rescher (1996, 2000) and Seibt (1996a, b, c, 2004, 2007, 2008, 2010), despite the fact that Rescher takes processes to be particular while Seibt takes them to be non-particular (general, possibly multiply-occurrent entities). I follow Seibt, since processes described by fundamental physical models oftentimes cannot be interpreted as particulars, as I will show.
[3] Borrowing the language of occurrents (processes and events) vs. continuants (things and structures and states) from Johnson (1921).

terpretations of various models[4] can be grounded in a general argument for the strength of process ontology within scientific theory and modeling. I call this argument the continuity argument.

The most basic formulation of the continuity argument is as follows:

(P1) All experiments/observations are event-wholes.
(P2) All experiments/observations contain at least one processual part: the act of observing.
(P3) The act of observing cannot count as an observation of an external world unless this act is the dynamic response to something in the world (condition of continuity).
(C1) Therefore, the portion of an experiment/observation that is not the act of observation itself must at least be actively dynamically potent, i.e., have a processual element.[5]
(C2) Therefore we know that there are processes in the world: the act of observing the world and the dynamics in the world that enable this act.

In what follows, I will take this basic formulation and show how it can be fully justified and made more precise. Of particular import will be the precisification of premise (P3), the condition of continuity. The successive precisification of this argument will produce three different arguments that represent three corresponding levels of commitment to real processes: (1) commitment to some process, (2) commitment to some process in the world, and (3) commitment to some process that can be described and codified in models and theories of worldly systems.

I begin the chapter with a brief introduction to process ontology, its history, and the relevant features of the ontological category "process" that I will be mak-

---

[4] For a sample from across the sciences, see Barwich (2018), Chen (2018), Dupré (2014, 2018), Earley (2008a, b, c, 2012, 2016), Ferner and Pradeu (2017), Finkelstein (1996, 2008), Guay and Pradeu (2015), Hartman (2005), Jungerman (2008), Kaiser (2018), Malin (2008), Meincke (2018a, b), Pemberton (2018), Pradeu (2018), Riffert (2008), Stapp (2008), Tanaka (2008). Of these authors, Joseph Earley, John Dupré, David Finkelstein, and Marie Kaiser are perhaps most representative of explicit moves toward strictly process *ontologies* within science, while the others tend to offer arguments that are process-ontology-adjacent or suggestive.

[5] This premise needs to be further clarified. An ontology that includes causal powers or potencies might argue that passive powers can explain how a system can engender an act of observation. However, this would require that we posit the existence of a property that is only knowable when it is destroyed: we only know the passive power in virtue of making it active. The arguments below are predicated on the epistemological assumption that, for scientific realism to work, we must only adopt ontological entities that are in principle knowable through experiments. Active powers are knowable in this way, but are nothing more than actual activities. Passive powers are not. This argument is presented in more detail in Chapter 2.

ing use of (§1.2). My aim is not to add to the ontology itself, but to provide reason to suppose that scientific modeling and experimenting practices necessitate this ontology. I then turn to a thought experiment: the unchanging room (§1.3). This thought experiment will provide us with the first and most basic version of the continuity argument, and will prime our intuitions for why we must suppose experiments and observations in actual science are processual. I then present the continuity argument in detail (§1.4), first as a simple justification for the belief that there are real processes independent of our acts of observation and intervention (§1.4.1), then as a complex argument for why we must suppose that our models must describe real processes if they hope to be empirically adequate (§1.4.2). I then conclude by noting that the result of this chapter does not preclude that things—those entities described in "substance ontology," like essences, objects, souls, stuffs, and so on—are real. The refutation of the claim that in science we are necessarily committed to things will come in Chapter 2.

1.2 Defining "Process" for Use in the Philosophy of Science

Our first order of business is to articulate an operating notion of "process." This notion is born out of examples, primarily. Processes are entities like:
(1) Motions: the motion of the earth around the sun, the vibration of air in a sound wave, etc.
(2) Interactions: the electromagnetic repulsion of one electron from another, atomic spectral absorption and emission, etc.
(3) Growth/Excitation: the development of a tree during its natal life cycle, the expansion of a gas, etc.
(4) Decay: the decomposition of a dead tree, the energetic fluctuation of a neutron in beta decay, etc.

However, these examples are merely intuitive, meant to establish that processes can be recognized in the world independent of a detailed ontological analysis. Importantly, I will not be adding to the existing analysis of the ontology of processes. Instead, I will be committing to one ontological analysis—Johanna Seibt's General Process Theory (GPT)—and provide the necessary additional considerations for the GPT to enable its application to scientific models in §1.4. In particular, I will be adding historical context and a translation scheme from ontological to scientific language.

1.2.1 Processes vs. Things

Briefly, the history of ontology contains two paradigmatic projects: the processist project and the substance (or thing) project. These two projects are characterized by different emphasis on the core epistemology of philosophical analysis; different prioritization of dynamics vs. statics as explainers, identifiers, and definers; and various secondary differences such as the primacy of theoretical definition over empirical practice and the like. The projects can be summarized roughly as follows:

(*Process*): Our knowledge of the nature of the world stems from induction over dynamic sensations and experiences and the recognition of similarities between changes and evolutions of the world. This means that the primary/fundamental ontological entity is dynamic, and we recover stabilities, structures, spatial relations, and static qualities/quantities as emergent features of systems of dynamics. To do this, we emphasize epistemological practices over theories.

(*Thing/Substance*): Our knowledge of the nature of the world stems from deduction from first principles and the recognition of static (or sufficiently stable), generalizable patterns that abstract away the inconstant aspects of the world we observe. This means that the primary/fundamental ontological entity is static, and we recover change, cyclic systems, temporal relations, and dynamic qualities/quantities as emergent features of collections of statics. To do this, we emphasize definition and theories over practice.

Those familiar with the history of philosophy will likely recognize that these two projects as I have glossed them are the work of various historical figures. Of particular import are Heraclitus and Parmenides, in whose work the dichotomy between the projects is most apparent, even if Parmenides was not committed to a substance ontology, denying not only change but also plurality. Heraclitus emphasizes that knowledge of the world comes from experiences, and so the primary entities of the world are dynamics. His *Logos* is the pattern of equivalent exchange found in the system of dynamic flows and motions, the "coming into being and going out of existence" of the world's "ever-living fire."[6] Parmenides, in contrast, denies that change is real, emphasizing instead that knowledge is of absolute and eternal truths, from which we deduce true claims about the world.[7] The resulting ontology is found as one interpretation of Platonic ontology: a primitive ontol-

---

[6] Heraclitus, fragment 30, translated in Kirk, Raven, and Schofield (1957). See also Plato, *Cratylus* 402a, for the apparent first instance of the attribution to Heraclitus of the claim that "one cannot step into the same river twice."

[7] Parmenides' proem "On Nature," translated in Kirk, Raven, and Schofield (1957, 243–253).

ogy of absolute forms from which are built the various combinations of qualities and their changes we observe in the world.

One can find thorough accounts of the histories of these two projects in existing works. For the history of the process project, see Clayton (2008), Eastman and Keeton (2008, preface), Rescher (1996, 2000), and Seibt (2017). For the substance project, see Moore (2012),[8] Robinson (2018), Seibt (1990), (and in general most historical analyses of metaphysics). I will not attempt to recreate such narratives. Instead, I only want to add a few additional historical points for those wishing to observe this dichotomy in the history of science.

We also see the opposition between process and substance/thing thought at various points in the history of science, and between various culturally divergent scientific practices. Three, in particular, are worth mentioning for the stark manner in which they paint the differences between these ontologies:
(1) Galenic vs. Han-era Chinese medical theory
(2) Roman, Islamic, and Scholastic vs. Mohist/Taoist physics
(3) The ontologoy of classical physics vs. that of quantum mechanics in 20th-century science

In each of these, we see different aspects of the applied dichotomy between process and substance thought. In (1), we see the difference between the anatomical definition of parts in Galenic medicine and the emphasis on whole-functions (literal "flows" or Qi) of the body in Chinese medicine.[9] In (2), the primacy of motions in the Mohist/Taoist physics contrasts with the primacy of bodies in (much of) Roman, Islamic, and Scholastic physics.[10] (3) represents the strongest historical contrast, and the quantum revolution is often taken as a motivation toward the process enterprise.[11] In the words of David Finkelstein, "Classically, knowledge is a mental representation of things as they are. An ideal observation informs us about its object completely and without changing it. ... [but] in a quantum epistemology, knowledge is a record or reenactment of actions upon the system" (1996,

---

8 Note that Moore presents his (very thorough) history of substance metaphysics as merely a history of metaphysics. Aspects of the contrast to process metaphysics can be found if one knows where to look—for example, in Chapter 18 on Heidegger—but otherwise the discussion clearly treats "things" (variations on substance) as the primary explanans of metaphysics.
9 See Kuriyama (1999). Note that Yuasa (1987) develops an account of body (both medical and philosophical) that trades on the same dichotomy noted in Kuriyama, and with different historical analysis. To some extent, the dichotomy between European and Chinese-Japanese approaches to personhood (including medical features) is also presented in Watsuji's (1988 [1935]) work "Fūdo" ("wind and earth" sometimes translated as "Climate and Culture").
10 See, for instance, Needham (1969, chs. 4, 7).
11 See Eastman and Keeton (2008), Seibt (2017).

18). In other words, knowledge in the quantum realm is not about classical definite properties or states of being, but rather about the activities of the experimenter and the dynamic responses of the system to those activities: processes, not things.

1.2.2 What Is a Process?

We turn now to our operating definition of "process." Following the Mohist/Taoist tradition, I treat "process" as a primitive ontological category. However, as usual in analytic ontology, we can further characterize this category by means of familiar category features, i.e., features which specify how an entity enters into linguistic, spatial, causal, and temporal relations. These are discussed in great detail in Seibt's work (1990, 1995, 1997, 2006, 2008, especially 2010, 2015, 2018). I summarize them here.[12]

(1) *Processes are general, not particular, entities:* they are individual in that they can be named and have features attributed to them, but they are not particular. I.e., processes are not inherently localized either in a single spatiotemporal location or in a single entity by necessity.[13]

(2) *Processes are subjectless:*
  (a) They are not alterations or modifications of things, or alternatively,
  (b) Their existence or occurrence is not dependent on something in which they occur.

(3) *Processes are occurrent, not continuant:*[14] They are temporally extended, and cannot be identified at a moment in time (they are not instantaneous).

(4) *Processes are not countable, but are measurable:*[15] One quantifies processes into amounts, which may be counted (10 joules of kinetic energy, for example, is a measure of the function of a thermal process equivalent to 10 processes of 1-gram-of-water heating by 1 K). Quantities attributed to processes cannot enable a mapping from a set formed of processes to the natural numbers. How-

---

[12] I omit the features described in the GPT that I will not make use of. In particular, Seibt points out that some processes can be determinately located in space and time because of their high degree of specificity, while other processes can only be indeterminately located. This means that processes are not necessarily determinately located, but can be, in contrast with things.

[13] This is similar to the non-particular ontologies of Sellars' (1952) and Leibniz. See, for instance, Rescher (1967).

[14] C.f., Johnson (1921), Johnston (1984, 1987b), Simons and Melia (2000), Seibt (2008).

[15] Note that we can count kinds of processes (e.g., excitations), but there are no "countable process individuals" (e.g., "the excitation of the neutron in this hydrogen isotope").

ever, it is possible to model systems of processes in such a way as to produce countable numbers of comparisons of processes.
(5) *Processes are not necessarily determinate, but are determinable:* This is a generalization of (4) above on a scientifically grounded understanding of the determinate determinable distinction.
(6) *Processes are individuated contextually:* The role of a process in a system, the dynamic responses to and perturbations of the process, and other functional considerations serve to characterize the process in terms of its context.
(7) *Processes are not changes, nor do they have temporal phases or stages:* rather, we say that processes can have their function measured by, or be partitioned into changes, phases, or stages.[16]

Of these, (1) and (2) will only become important when we move to the later chapters, especially Chapter 2. (3) is essential for the continuity argument to come. This is because the inability to find instantaneous properties of interventions or observations will indicate to us both that interventions and observations are themselves processual, and that the systems we intervene on similarly lack instantaneous properties. (4) will prove particularly important when we reach Chapter 4. (5) will largely play only a background role, but is worth mentioning. (6) is essential to this chapter, and will be discussed in great detail within the context of scientific experiment in section 1.4.2.3. (7) represents a linguistic point. Together, these seven features (plus the one I have omitted) define the category features of process slightly modified from the General Process Theory (GPT) ontology, defined and defended by Seibt.[17]

Importantly, the features of processes are the means by which we identify them in the world. We also need to construct a means of classifying processes into types to meet different linguistic, descriptive needs. We can construct these classifications by noting not how we identify processes, but how we differentiate

---

[16] This is a slight departure from Seibt's GPT, but is important for the purposes of this work. Namely, we must eschew describing processes as if they are composed of sequences of states. States, as terms within our physical models, can act as designators of processes (in that we can collect contextual information about processes into mathematically defined states), but they cannot act as descriptors of those processes (the process is not built up from those states). Note here that "state" is used according to its common usage in the philosophy of science: a state is a collection of definite values for the relevant variables of a particular system's physical realization.
[17] In what follows, I simply commit to this ontological framework. There are a few additions to it I advocate in later sections and chapters, but for the most part, the GPT is an ontology of pure (subjectless) processes that meets the linguistic needs of the ontologist and the realist. I will show that this ontology of subjectless processes can be reconstructed and evidenced within scientific practice and theory.

them from each other.[18] These classifications will allow us to reconstruct every sentence involving things, substances, events, properties, actions, relations, structures, vehicles, and so on. They are:[19]

(1) *Participation:* The system(s) involved in the process. For example, two motions may differ in the types of system in which they occur, despite sharing all other features and classifications.

(2) *Process Structure:* The relations between processes or processual parts that obtain (cycles, sequences, etc.). For example, two quantum-field-theoretic interactions may differ solely in that they have a different number of closed loops in k-space (as represented in their respective Feynman diagrams).

(3) *Dynamic Context:* The connections between processes and the event-wholes within which they obtain. For example, two radiative emissions may differ in the scope of their influence on their environment (the sun vs. my sunlamp). Two processes may also differ according to the pragmatically chosen effects we consider. E.g., the radiation of the sun has physical, biological, ecological, and sociological effects. We would want to differentiate between these in labeling our processes.

(4) *Mereological Signature:* The non-transitive relations that processual wholes bear to their own parts, both functional and spatiotemporal.[20] For example, the performance of Rachmaninoff's 2nd Piano Concerto, the motion of a pendulum, and the activity of the sun shining differ in how alike the parts of these processes are to their respective wholes. The concerto has no parts that are alike to the whole, the motion of the pendulum has some parts that are alike to the whole, and any part of the sun shining is like the whole. Each of these examples displays a different manner in which, e.g., the functional

---

18 This, too, is drawn directly from Seibt's work. However, it is worth noting that Pemberton (2018) offers a somewhat similar analysis of how we individuate and classify processes in science. Many of Pemberton's categories end up overlapping with those described by Seibt, while a few do not (e.g., Pemberton admits a classification of processes in terms of "originating things"). I have purposefully committed to Seibt's account over Pemberton's because Pemberton's is not yet developed enough to meet the needs of a serious *pure* process realism in science. Indeed, I believe Pemberton's work ultimately fails to argue that processes are effective posits for explaining either linguistic or scientific data.

19 I omit the classification of "dynamic shape" (Seibt 2010, 49) because this seems to track linguistic differences between process descriptions, not physical differences necessarily. As such, this classification is important for normative projects that take linguistic data as their primary data, but will prove less interesting for the project here.

20 One can also specify other sorts of mereological relation, such as material relations.

features of a process distribute over spatiotemporal regions in which that process occurs.[21]

The mereological signature is the primary means of classifying processes qua reproducing all linguistic forms. Type-1 processes are activities like "it is raining" that are everywhere and everywhen like-parted. I.e., any functional part of the process is a spatiotemporal part in which the entirety of the process occurs. Type-2 processes are the processes of "stuff" like water or mud, marked by non-maximal and non-minimal spatial like-partedness and maximal temporal like-partedness. And so on.[22]

For the purposes of what is to come, the most important classificatory differences between processes come from (1), (2), and (3). This is because our goal will be to show that the entities or terms we find in our physical models differ according to differences in their dynamic context, structure, and participants. (4) will play some role as well, although mostly in the final analysis, when we eschew thing-claims in our models entirely in favor of process claims. At that point, it will be important to show that any apparent thing claim can be physically understood as a placeholder claim about purely processual entities. As a result, any statement involving apparent things in or produced by our physical models can be reproduced entirely in terms of the subjectless, general processes of the GPT.

---

[21] For more, see Seibt (2015). There is a slight difficulty here in that ultimately we will need to eliminate talk of spatiotemporal regions in order to maintain a process-ontological interpretation of physical theory. To do this, we will need to develop an independent means of discussing spatiotemporal regions as if they are merely emergent functional features of some fundamental interaction(s) and their concatenation. Finkelstein (1996) has already gone a long way in developing such an account. The modus operandi of such an account is as follows: the topology that forms spacetime geometry is defined by some fundamental basis for comparison of the strengths of a particular sort of interaction (or interactions). This ability to functionally identify these interactions then allows the construction of topological regions representing regions of similar-strength interaction and function, which in turns allows the definition of a differential manifold without the need for an underlying manifold of points or "locations." In turn, this alters our definition of mereological signature slightly in the context of physics specifically. That is, we now consider mereological signature as a generalization of the idea that we can describe relations between various purely functional decompositions of processes and how they distribute over other such decompositions. One of the relations we can describe will be the way in which certain functional parts of processes distribute over spatiotemporal parts of the region in which a process occurs. However, we will understand the spacetime parts of the regions in which the process occurs as themselves merely functional parts of processes relating to how they act and are acted upon by the fundamental interactions from which we build our spacetime topology.

[22] For full details, see Seibt (2010). I do not reproduce these details here because they will serve little purpose for the coming discussion.

1.3 A Thought Experiment: First-Level Process-Realist Commitment

With an operating notion of process in place, we can now turn to the arguments for why we are justified in committing to processes as part of the ontology of scientific models and theories. As stated in the introduction, this argument begins with a simple version and proceeds through more complex iterations further on. Our first argument for the reality of processes is that processes must exist in order for observation, experiment, and scientific knowledge to be possible. This establishes only that processes should feature somewhere in our ontological understanding of science, not that we must accept any specific processes or find real processes in any specific place. To see this, we turn to a simple thought experiment.

1.3.1 The Unchanging Room

Imagine a room in which nothing changes. Imagine we sit in this room and we imagine attempting to understand this room scientifically. Surely, to do so, we will first need to observe the room, and then perhaps to experiment on it. Perhaps we might even wonder if there is anything we can infer about this room or anything in it. To that end, we must ask ourselves: what can we observe in, experiment on, or infer about this room? The simple answer is: absolutely nothing.

Let us consider examples. I have stated that we are in a room. Can we observe that there are walls around us? We certainly cannot see them. Seeing walls would require propagating light reflecting off of the walls. Seeing would require a dynamic response in our eye in response to the light. The recognition of this as an observation of the wall would require an activity within our nervous system in response to the dynamic response of our eye.[23] All of these are dynamics, which will be char-

---

[23] For those curious, the literature on observation nowhere disputes that observation involves perceptual processes as I have stated. E.g., though Hempel and Feyerabend both dispute the naive account of observation in which observation is just a perceptual process, it is because they believe respectively that perceptual processes must originate in determinate facts about objects and things (Hempel 1952, 674), and that perceptual processes must include processes in measurement tools and apparatuses (Feyerabend 1985 [1969], 132–137). Helmholtz similarly disputed the naive account on the basis that perceptual processes could not register changes in a system smaller than some human limit, and so had to include as observations artificial processes, or "artificial methods of observation" (see the analysis of Olesko and Holmes (1994, 84)). Thus, while the theory-ladenness of observation, the distinction between observation and experiment, or even whether

acterized by measurable changes, phases, and stages (cf. Feature (7) of processes above), but the room contains no change.

We might think that we could somehow be in constant contact with the wall. Perhaps there is some static relationship or link between us and the wall, born physically by some stationary light wave or electromagnetic structure. Even supposing this (outrageously incorrect physical assumption), we would still be incapable of observing the wall. We would be greeted with at most one undifferentiated and unchanging image. How could we notice that there is a wall in front of us without comparison, without shifts of focus and active differences in our response to one part of the image or the other? Past experience cannot be used for comparison. The thoughts through which we could compare the image to past experience are, after all, activities (type-1 processes). Activities are processes, and so should have associated measurable changes. But the room contains no change.

What about something other than sight? Touch and hearing might work where sight fails. You certainly cannot touch the walls. Touching them would require that you move your arms, which cannot occur without change in your position. Even granting that you could already be in contact with the wall, the feeling of touch requires that there is an electromagnetic interaction between you and the wall, i.e., a process of equal and opposite energetic exchange through repulsion. But the room contains no change. Even further, you would not know that you were in contact with a wall as opposed to anything else unless there was an appreciable difference in the pressure you feel in your fingers as your hand presses against the wall and against something else. Measuring such a difference in pressure would require multiple sensations, or a comparison to past experiences of sensations, which we have already ruled out. Hearing is similarly impossible, since we hear because of vibrational changes in air pressure causing oscillations of the cilia in our inner ear. But the room contains no change.

Perhaps observing the walls is impossible. Can we experiment? Surely that too is impossible in the absence of the ability to observe. Moreover, experiments are typically differentiated from bare observations by the observer's ability to manipulate and intervene on the system being observed.[24] Interventions are activities.

---

observations are *of* data or phenomena (Bogen and Woodward 1989), the processual character of the perceptual processes involved in observing is never in doubt.

24 See, for instance, Cartwright (2001, 2002, 2006), Hausman and Woodward (1999, 2004), Hitchcock (2006, 2007a, b), Suárez (2013), and Woodward (2003, 2014a, b, 2015, 2019). Interventionists of the causal-Beyes'-net variety often remark that interventions are represented as "an exogenous variable ... with two states (on/off) and a single arrow into the variable [the on-off switch] manipulates" (Eberhardt and Scheines 2006). This means that the intervention is understood implicitly as the causal action of the external manipulator on the internal system variable. Although I think

Manipulating a system requires us to dynamically alter it in some way. They require change in the room. But the room contains no change.

What about inferring that the walls exist? We might imagine that, even barring an ability to access the walls scientifically, some manner of abstract, first-principles analysis might allow us to deduce that there are walls. Perhaps we might infer that a benevolent, non-deceiving deity would not place us in a room we could know nothing about. Indeed we might so infer, were we capable of inferring at all. Inference would require us to engage in an activity of mind, perhaps manifest as neurochemical activities in our physical body. These would entail the existence of measurable changes, but the room contains no change.

At best, we might hope that we simply know that there are walls without justification or inference, and that this knowledge is somehow manifest as a static property of our minds. Such knowledge would look rather different from our everyday knowledge, however, since it would be inutterable, inaccessible, and unresponsive to processes of thought or consideration or the like. You could not deduce from it, nor induce over it, nor reevaluate it, nor build a theory or model or concept from it, or anything else. Perhaps such knowledge could exist, but it certainly wouldn't count as scientific knowledge.

So the walls are beyond our ken. Similarly, every other so-called concrete or stable entity in the room would be inaccessible. Might there be something in the room that is observable or inferrable independent of the physics of the room? As an example, we might appeal to Kantian forms of intuition: space and time. Perhaps, in this room, we may be incapable of observing anything and yet capable of having observations were we only permitted to engage in that process. Perhaps, you could know that there is time or space simply in virtue of their necessary existence for your metaphysical and physical state of being.

Once more, however, we are defeated. The Kantian forms of intuition are justified on the basis of experience. Critically, they are justified because we have multiple and dynamic experiences of the world. The recognition of passing time comes from recognizing that there are indeed different moments to be ordered in a sequence. If we have, at best, a single static image of the world with which to judge, we cannot possibly recognize ordered moments from this single state. Similarly, the recognition of space is born from our ability to order sensations in terms of nearness and farness. Again, if we have but one static image with which to judge

---

the talk of manipulating variables (predominantly influenced by Woodward (1999, 2003, 2004)) does not track the ontology of experiment (variables should not be treated as real entities), this representation is still useful for revealing that, indeed, interventions depend on an activity or process.

(and probably not even that), we will be incapable of even recognizing any difference in the nearness and farness of the impressions in that image.

We have one final recourse: appeal to the most unassailable of ideas. For instance, can we infer that we exist in this room? If we take Descartes seriously, we only know this because we think. However, thinking is an activity. We cannot think statically. Thinking and thereby knowing that we exist require change. But the room contains no change.

1.3.2 Without Process, No Observation

The problem we face in this room is general. Observation, experiment, and inference are all processual: type-1 or type-3 processes on Seibt's account above,[25] occurrents in Johnson (1921) and the literature on endurance and perdurance, activities in Vendler's (1957) linguistic account, causal arrows in interventionism, etc. Observations, experiments, and inferences are all identified by involving measurable changes, and by being occurrents (cf. §1.2.2 above). Our mistake was to suppose that these processes could exist in a room where dynamics are prohibited.

However, this is essentially an admission that without processes, we cannot have observations, experiments, or inferences. Given that each of these are essential to the practice, development, and improvement of science, this means that science is impossible without processes. Without processes, we cannot have observations, and without observations, science is empty. Theories and models cannot be built independent of experience, and experience is processual.

This matches how philosophers of science understand observation. Carnap and Feyerabend, and more recently Kuby (2018), all put it rather well:

> Rain-observing ... can perhaps be characterized as being found in certain conditions (namely when it is raining) or if rainlike audible or visible processes are present, and the eyes or ears of B are in the appropriate relative position to these processes) and as stimulating such and such observable bodily reactions ... (Carnap 1987, 460)

> Now it is most important to realize that the characterization of observation statements implicit in the above quotations is a causal characterization, or if one wants to use more recent terminology, a pragmatic characterization ... (Feyerabend 1962, 36)

Kuby argues that both of these interpretations of observation statements are to be understood as statements about causal processes (see, e.g., 2018, 12). Indeed, we are meant to understand observations as "protocols" for interacting with and learning

---

25 That is, either activities or accomplishments/developments respectively.

about changes in the environment and physical state of the observer, according to Kuby.

Indeed, the processual character of observation is never under dispute in the literature. There are disputes about (a) whether there is an object of observation (the system observed), (b) whether the thing observed is evidence or data or phenomenon, (c) what can be inferred from observations (but not *observation*), (d) whether observations can be treated as independent from existing concepts or theories (theory-ladenness), etc.

Drawing on Carnap's and Feyerabend's theories, we can put the points above more pointedly. Scientific knowledge is, most basically, built out of collections of experiments and observations that together enable the construction of some model or theory codifying and generalizing the set-up and outcome of each of those experiments and observations. So, if scientific knowledge depends on and is about observation, and observation cannot exist without processes, then our scientific knowledge is dependent on and about processes.

This is the most simple argument in favor of process realism. By showing that processes are essential parts of observation, we can infer that our models and theories in science must at least contain reference to processes as real entities in some capacity. The most obvious way in which they do refer to process is whenever they refer to or tacitly assume the truth of one or more observation sentences. In other words, I always know that processes are real in science because scientific models and theories expect me to be able to observe. The act of observing is the first and most basic process to which I must commit in science.

In addition to this, we learn three more refined points with which we will construct the second and third iterations of the argument for process realism:
(1) Our ability to observe is dependent on two processes:
    (a) An external or initiating process, a change in environment and an interaction between that environment and the observer
    (b) An internal or responsive process, an activity in the observer
(2) There are three meanings of observation, only one of which is essential to the process of science. These are:
    (a) The facts, features, or entities that ground the truth of an observation claim; the direct object of the act of observation (as in "I observed beta decay" whenever a real beta decay process occurs)
    (b) The statement of an observation itself (as in the dialectic exchange "There are real processes," and "that is a good observation")
    (c) The act of observing (as in "I observed alpha decay by *using* a geiger counter")

(3) Observation (2c) is identical to the responsive internal dynamics. I.e., observation is not just dependent on the existence of these dynamics, it *is* these dynamics.
(4) Observation (2a) is at least dependent on the existence of the external initiating processes, although it may not be completely processual (there may be real things being observed, so long as they are admitted to be dynamically potent).
(5) Any successful observation will necessarily posit the inseparability of the first and the second processes, 1a and 1b.

In other words, for every observation you claim to have made, it is possible to provide the necessary processes that initiate and constitute the observation.

Notice, finally, that it is not the lack of a subject that makes the unchanging room unobservable. In fact, the recognition that there is a subject observing is itself an observation we cannot make in the unchanging room. This should indicate to us that the only essential feature of our observations is that they are processes qua general, non-particular, subjectless dynamic activities, not that they are activities *of* anything at all. Whether or not these activities have a subject is superfluous. However, one is welcome to assume, for now, that the activity of observation must have a subject. I will refute this and all other underlier arguments in science in the next chapter.

1.4 The Continuity Argument

It is often assumed that the antirealist has better claim to the idea of empirical adequacy than does the realist. The realist "goes beyond" empirical adequacy in making their realist claims. Indeed, against most realist positions, the antirealist has a point: the realism being posited is not merely a means of enabling the successful description of pragmatic aspects and outcomes of experiments. Instead, the realist seems to posit their realist claims to ground empirical adequacy of a model in some metaphysical feature of the world. The antirealist, then, seems to commit to far less than the realist, and so they do not fall victim to metaphysical extravagance.

However, the thought experiment above shows something different. Crucially, we must ask if there are any features of the world without which empirical adequacy would be impossible. I.e., what happens when we seek not to explain empirical adequacy in terms of the metaphysics that enables it, but instead seek to understand the metaphysical claim contained within the claim of empirical adequacy? The answer is, we get an argument for process realism, and the subject of this section. Namely, we get the continuity argument.

The program for this section is as follows. We begin with a simple version of the continuity argument (§1.4.1). This will establish that we are justified in inferring that there are real processes external to the observer, i.e., the second level of realist commitment. Having presented the simplified version of the argument, we turn to a defense of the full-fledged continuity argument (§1.4.2). This argument shows that we are also justified in inferring the existence of real processes as described by our models and theories, i.e., the final and most important level of realist commitment for scientific realism. Along the way, we will show how and in what way we partition and identify processes within our experiments, and use this to show that the entities we infer from our theories, models, and experiments fit the description of subjectless activities in the GPT (see §1.2).

1.4.1 The Simple Continuity Argument

In the thought experiment above, I showed that for every observation or experiment, there must exist some process in order for that observation or experiment to occur. I could not, with this argument alone and with this most basic understanding of observation, learn anything about the type or features of those processes, nor indeed whether those processes were processes in the experimental system. The next step in our argument for process realism is therefore to further specify the general features of experiments and observations, so that we may see what processes must exist in our experimental systems, and what features they must have.

The simple continuity argument is as follows:
(P1) Observations of an external world are occurrences.
(P2) Such observations consist of two temporally extended parts: (1) the activity of the observer alone which is a dynamic response to (2) whatever exists or goes on outside the observer in the system being observed.
(P3) (1) and (2) are dynamically continuous: any distinction between them will need to track only pragmatic or partial physical distinctions between them, not ontological distinctions.
(C1) Therefore, at least one part of (2) must be a process (since if it were not, there would need to be a determinate and absolute ontological distinction to be drawn between (1) and (2)).
(C2) Therefore, there exists at least one process of indeterminate type external to the observer that is responsible for the activity (type-1 process) of the observer in their act of observation.

This simple improvement on the argument for process realism in section 1.3 rests on two features of observation. The first is that the act of observing a system is indeed an observation of an external system. The second is that any criterion by which we identify the internal process will also be a definition of the external process. Our means of identifying the internal and external processes are identical where they are in spatiotemporal contact with each other. Moreover, this means of identifying the two processes, and partitioning them from each other, is itself dependent on some process to be specified contextually as per the classification of processes (3) in §1.2. This suggests that the internal and external processes are identified by features unique to subjectless processes, namely features (3) and (6), occurrent-ness and contextuality (§1.2). Thus, the processual nature of the act of observing a system necessitates that the external environment that enables this act is at least as processual as the act itself. Thus, processes exist external to observers, and therefore exist in experimental systems.

1.4.1.1 The Act of Observation Partitioned into Two Processual Parts

When we make observation statements, we typically specify both a system and an agent that engaged with the system. Examples include, "I observed a tree," "Curie observed the radioactivity of Uranium," "Kandinsky observed blue." However, we implicitly assign or assume two key features of such statements that determine the details of their specific real-world commitments. First, in each, we assume that for every such statement, we can answer the question of "how" the observation took place. I observed the tree by seeing it. Curie observed radioactivity with an electrometer that produced deflections of a needle in response to Uranium rays. Kandinsky observed blue by mixing paint and looking at it. Second, we assume that the subject and the object of the sentence are not identical. I am not the tree. Curie is not radioactivity, and Kandinsky is not blue.

We can express these two assumptions as follows:
(1) Essential to each observation is some dynamic act involving a physical perturbation of the observer, and
(2) Essential to each observation is the assumption that this dynamic act is the response to some external source.

(2) provides for the assumed difference between subject and object; notice that (2) is true even in cases of reflexive observation such as introspection. (1) provides for the answer to our "how" questions. We will always refer to some change in the observer's activities or their environment (usually both) to answer these "how" questions.

1.4.1.2 Two Parts, One Whole Activity

While we admit that observation acts have two parts, these parts are co-dependent. The identifying features of one are identifying features of the other as well. Put simply, an observer cannot have a response to some external system if the external system doesn't initiate this response. The external system is therefore partially defined by the features that identify the act of observing as an activity.

To see this, let us consider the act of seeing. We have already said that we assume that "I see X" entails both that I engaged in some internal perceptual process and that there is some entity that initiates this process. My act of seeing is identified by core processual features: it is non-instantaneous, context dependent, and cannot be counted but can be measured. Importantly, the context dependence necessitates that my act of seeing is identified by the dynamics involved in the context in which I see. I.e., my act of seeing is partly identified as a process by the fact that it is an active dynamic response to my environment.

In acts of seeing, this dynamic context takes the form of an electromagnetic flux in the eye. I physically see because my eye responds to changes in the electromagnetic environment of my eye. I.e., me seeing involves a necessary presumption of an electrodynamic interaction between my eye and the environment. It is this electrodynamic interaction that defines the dynamic context of my act of seeing, and to thereby identify this act of seeing as this particular act of seeing and not that one.

The reverse is true of the environment, experimental system, or external entity that we observe. The external entity qua *observed* entity only counts as observed when there is a dynamic response in some observer. The necessity of this dynamic response means that the observed entity will be partially identified as observed only when we can attribute to it some real dynamic effect. I.e., to be observed, the entity must cause an act of observation. This can only obtain if the observed entity has processual features. It must be non-instantaneous, since there can be no instantaneous response in the observer. It must be uncountable but measurable, since the response in the observer is uncountable but measurable. It must depend on the dynamic context of the observer(s), since how the observer responds will define the observed object (qua observed).

But this means that the external part of our observation, the system observed, is identifiable as a process. The same features that identify my act of seeing as an activity (or development), a type-1 or type-3 process, are the features that we use to (partially) identify the object of our observations. In particular, both are identified in observation sentences by a non-instantaneous, non-countable, actual dynamic interaction. This means that both the act of observing and the object of observation must be at least equally processual. There may be additional features that define the object of observation, and so allow it to contain other sorts of entities (e.g., things that undergo processes). Nevertheless, any observation—both the act of observing and the observed system—must be processual in character. Indeed, given the connection between the former and the latter, we should say that the two form a single dynamic event "the observation" that contains two processual parts.

1.4.1.3 Experiments Assume the Existence of External Processes

The upshot is this: the processual character of our observations requires that whatever we observe is at least as processual. In short, our reasoning pattern here is an inference from a known process (the act of observing) to an entity of unknown type (the external environment/experimental system). Since the two are necessarily co-identified by a set of processual features (non-instantaneity and dynamic contextuality), the unknown entity acquires these processual features in virtue of being connected to the known process. In other words, the fact that we observe a system entails that the system is dynamic, such that it can be observed. Were it not, our "observations" would be impossible—we would not be dynamically responding to anything. Unlike in §1.3, however, we now know that the something we are responding to, the external experimental system plus the surrounding environment, contain real processes because they have the identifying features of processes listed in §1.2. We are therefore justified in inferring that real processes exist in our experimental systems.

1.4.2 The Complex Continuity Argument

We now have a basic argument pattern for process realism, and for specific processes within our experiments. We begin with some whole event. We show that this event can be partitioned into parts that meet a special condition at the partition between them (non-instantaneous identification, measurability but not countability, co-identification). We show that one of the partitioned parts is a process (usually by feature analysis). We then show that the special condition entails that the other part is a process in virtue of being identified by the same processual features as the known process.

Those familiar with mathematics may notice that the special condition at the partition looks a lot like a continuity condition. E. g., in the Dedekind cut construction of the real numbers, the method involves first producing a partition in a set, then showing that both partitioned parts are co-defined at the partition and that the partition can be moved infinitesimally (this is the equivalent of the "non-instantaneity" of the partition between parts of an experiment). Our argument can therefore be written as:

(P1) Experiments are event-wholes (i.e., single events, possibly with legitimate compositional parts).
(P2) We know that these wholes have at least two processual parts: our act of observing (the perceptual process) and our act of intervention (the manipulation process).
(P3) Experimental wholes can be completely partitioned into parts defined by our acts of observing and of intervening.
(P4) All parts of the experimental whole not identical to the acts of observing and intervening have features defined by these acts in virtue of them being continuous (the condition of continuity). I.e., the physical definition of the partition between these parts is non-instantaneous and co-identifying.
(C1) Therefore, all parts of the experiment must be processual, since they have processual features.
(C2) Therefore, there are real processes within the experimental system that are describable using models that imply or refer to the observation and intervention acts of the relevant experiments.
(P5) In order for our models to be empirically adequate, they must contain reference to or implication of a class of observation and intervention acts.
(C) Therefore, our empirically adequate models must describe real processes.

Our development of this argument pattern into one capable of producing true realism (commitment level 3—(C) above) will involve making precise the continuity condition (P4) that was implicit in the basic argument, and making precise the means by which we generally identify parts of our experiments (P3). We begin with the former.

1.4.2.1 Defining Continuity

Continuity appears in philosophical literature as both premise and conclusion to many arguments. Roughly, there are four types of continuity that are used: functional, causal, spatiotemporal (topological), and conceptual. Common to all of these is the idea that for any continuous entity of the appropriate type (a function, a causal agent or client, a spatiotemporally extended entity, or a concept), (a) we can partition the entity into parts, (b) we know that for any such partition there is some sense in which the partition is co-identifying of the two parts. We might reverse this and say that for any two entities said to be continuous with each other, we know that there is some co-identifying boundary between the two without which the two cannot be identified.

For example, in the case of a spatially continuous entity, we would say that two partitioned parts of the entity are defined as parts by a shared boundary, and that this boundary is the sole means of defining them as parts. In other words, the two spatially continuous parts cannot be identified as separate entities, given that they share at least one necessary feature in their shared boundary.

We could make this more precise by speaking of spatial cuts of the continuous whole defining sequences along spatial dimensions, and noting that any sequence of such cuts will necessarily converge to another cut of the entity.[26] In this way, we would say that the entity lacks "gaps" or "breaks." If such a gap were present, a sequence of cuts could be constructed that would converge to the gap.

This simple definition of continuity is enough to define it for our use. Essentially, we will be generalizing the mathematical notion of continuity to enable talk of continuous parts of a single general entity and continuous pairs of general entities. Namely:

(*Entity-Entity Continuity*): Two entities are continuous with each other if for any total identification of one, that identification also acts as a simultaneous partial identification of the other in the same context of identification.

And equivalently:

(*Part-Whole Continuity*): Two entities are continuous whenever there is some third entity of which both are parts, for which both completely compose the whole, and such that all three are the same metaphysical type.[27]

In short, we define continuous entities by their lack of gaps. If one entity ends (along some considered dimension), that end must define the beginning of the other.

This definition has a few key features that will play an important role in the discussion to come:

(1) *The definition is entity-type independent.* Importantly, we need not specify what sort of entity is being evaluated, so long as we specify how we identify the en-

---

26 This is the "complete convergence" condition of the real number line that acts as a differentiator from the rationals.
27 This is the less-obvious version of the continuity condition. One might say that atoms in molecules or the various parts of an organism are continuous under this criterion, which would be dubious claims. However, in both instances, the criterion fails to apply (at least on orthodox interpretations of all the relevant entities). Importantly, the atoms in a molecule, under orthodox thing-ontological or substance-ontological interpretations *do not* completely compose a molecule on their own. Instead, they require some third non-substantial entity—the relation between atoms or structure of the molecule—to form a mereological whole. Since this third entity is not the same ontological type as the atoms (on thing interpretations of atoms), this interpretation does not meet the continuity condition. However, I think it is natural and intuitive to jump to this conclusion, at least initially.

tity and are consistent in this identification. So, for instance, metaphysical substances can be continuous,[28] so long as we know how to identify them. Therefore, the recognition of continuity does not presuppose any particular ontological posit, which is essential.

(2) *The definition entails the corollary that no two entities may be continuous if they are not of the same metaphysical type.* This follows from the requirement that both entities can be either identified by the same means (e.g., with the same properties or by the same methods) or that together they completely compose a context-sensitive whole. This is essential to any definition of continuity, since implicit in the idea of continuity is that the continuous entities flow into each other in some definable manner. E.g., if I move my hand in a circular pattern, part of the motion of my hand flows into another part of the motion continuously, and we recognize this because, e.g., the flux of kinetic energy in the region can be defined with a continuous function.

(3) *The definition does not entail that two continuous entities need to be the same physical type.* For instance, we may say that a freshwater river and the ocean are continuous with each other, despite being different physical mixtures of water and saline. We might also say that the motion of my hand is continuous with (or is a continuous part of) the motion of the door I open with it, even though they are physically distinguishable by, e.g., their electromagnetic or thermodynamic signatures.

(4) *The definition can be used to claim that two entities are continuous if we can do one of the following:*
  (a) Find an identification method for one entity that also partially identifies the other. This is because any total identification of the first entity must include the identification that acts as the partial identification of the other. The definition is weak enough to allow that two systems can be continuous in one sense or experimental context while discontinuous in another.
  (b) Find a system that is a whole composed out of the two entities. For this, we will need to show that the whole is of the same metaphysical type as the two entities. This is to preclude the so-called unrestricted composition mereology from making continuity an empty concept.

(5) *The definition allows us to say that two systems are metaphysically continuous while also saying that the systems can be physically isolated with the right methods.* This is critically important for the practice of physics, at the least.

---

[28] It is worth noting that metaphysical substances are not primarily understood to be "stuffs" like mud, but are usually understood as primarily a thing-like entity (and these are discontinuous).

One of the tasks of the later sections is to show that the metaphysical continuity we locate in two continuous systems tracks exactly the physical continuity of their respective processes as described in our models. We do not need to show, however, that two systems are totally continuous in order to show that some physical continuity between systems appears as an essential feature of models of the system, and therefore is a necessary process-metaphysical posit of the model.

With this definition in hand, we can now turn to the most complex and specific version of the continuity argument. The definition will become relevant in §1.4.2.4.

1.4.2.2 Experiments Are Wholes

In this section, I merely wish to defend that experiments are wholes in the first place, independent of what type of entity these wholes are. The argument is exceedingly simple: experiments are wholes because we can name them as singlet events. Here is Thomson's experiment, here is Pauli's, here is Hypatia's, here is Curie's, here is Franklin's. We can also sort them into types, if we so wish: a Thomson Scattering experiment and a Franklin Spectroscopic experiment are identifiable and reidentifiable. E. g., Thomson's Scattering experiments involve scattering and bombardment and electromagnetic interaction making use of particular tools and set ups. When I see these features in place—the tools being used properly, the interactions and observable changes occurring normally—I recognize a Thomson experiment.

While I take it as uncontroversial that experiments are wholes, it is worth noting that they are not wholes of necessarily physical substances. Experiments are not identified in the same way as a statue: as composed of a particular form and matter. Even experiments in chemistry, plausibly experiments on particular types of matter, are not defined by those substances but rather by the observations and interventions performed on them. We may say that the experiment essentially involves the particular chemical substance being studied, but we would not say the experiment itself is that substance, even in part.

The upshot is this: we must keep in mind that, if we wish to support realist claims on the basis of the existence and features of experimentation, we must ensure that our inference patterns are appropriately independent of the particulars of any single experiment. For this reason, we rule out that experiments can be anything other than events, and we seek to ground our inferences to realist claims in only the mereological, metaphysical, and conceptual features of this special type of event.

1.4.2.3 Partitioning the Experimental Whole

I now argue that experiments can indeed be partitioned into parts. This partitioning is the result of differentiating the act of observation, the system dynamics and the initial dynamics by means of perturbative dynamics, i.e., intervention acts. In partitioning the experimental whole into these parts, we define the scope and purpose of our models, the description of the experimental system. This means that our partitions of experiments into parts (intervention, system, outcome observation) are non-instantaneous and co-defining.

Let us consider a simple example to develop this. Consider the experiments performed by Rutherford and his doctoral students on the scattering of alpha particles fired through sheets of gold foil. In this experiment, an alpha-decaying substance is prepared in a lead block with a small aperture so that the emitted alpha particles will be fired in a single, controlled direction at long intervals between emissions. These alpha particles are then fired toward a thin sheet of gold foil, behind which sits a fluorescent screen. We observe that, corresponding to each emission event, the fluorescent screen flashes with light in a small region on the screen. By treating the alpha particles as being emitted unidirectionally, we can then use the flashing point on the screen to measure the deflection of the alpha particle from its original emission trajectory. We then infer that, since the alpha particle will not interact electromagnetically with anything other than the gold sheet, that the deflection is the result of the alpha particle passing through the screen. By repeating this many times, we can use statistical analysis to calculate the approximate size of the deflection region within the gold sheet, since greater deflection will correspond to a greater electromagnetic interaction between the alpha particle and the gold sheet. Rutherford then used this result to posit (incorrectly) that this strength of deflection corresponds to the size of the atom's positive charged substance, and to construct a model of the atom now known colloquially as the "plum pudding" model.

This experiment illustrates the means by which we isolate and differentiate the parts of an experiment. The experiment as a whole is a collection and statistical analysis of the temporally distinct events of an alpha particle following a particular trajectory. We might call each of these events "singlet experiments." Each singlet experiment consists of a single continuous electromagnetic flux:
(1) The alpha particle (carrying a $2e^+$ charge) is fired out of the lead box,
(2) follows its trajectory up until the gold sheet,
(3) has this trajectory deflected through electromagnetic interaction with the gold sheet,
(4) flows through the new trajectory,
(5) then electromagnetically interacts with the fluorescent screen to produce a flash of light (electromagnetic radiation),

(6) that causes our eyes to respond to and register the dynamic end of the process.

We immediately recognize that the experiment consists of this single process of electromagnetic flux, and linguistically codify this recognition by referring to the process as "the trajectory of the alpha particle." However, we also recognize that we must impose divisions on the singlet experiment as a whole in order to appropriately model the process, and potentially to infer something interesting about the phenomena to which we do not have direct access. In this case, we wish to use this process as a means to measure the average strength and size of the deflection of the alpha particles, so that we can learn something about the electromagnetic properties of the gold sheet. We therefore divide the system into three parts when we go to model the experiment:
(1) The initial trajectory of the alpha particle
(2) The final trajectory of the alpha particle, including the flash of light on the screen and that light's propagation into our eye that we take as evidence of the final direction of the trajectory
(3) The interaction between the gold sheet and the alpha particle, which is assumed to take place over a negligibly small (but non-zero!) period of time for the sake of simplicity in modeling

Now, we bring in counterfactuals. If it were not for the gold sheet, (1) and (2) would be identical.[29] Therefore, (3) is the salient dynamic link between initial and final trajectories. Since we believe that this interaction must be electromagnetic, we then model (3) as a deflection of the trajectory caused by proximity to an electromagnetic source and the particle. From there, the model is relatively simple to construct.

What we have done here is divide what was once a single continuous event— the singlet electromagnetic flux from preparation to perception—into three parts. One of the parts—(1)—is uninteresting, since it represents the dynamic origins of the dynamics of interest. Another part—(2)—is interesting only insofar as it enables our analysis: it is the dynamics of the system that we actually observe directly. The last part is partitioned from the other two by the assumed relevant dynamics of the system we are studying. Namely, we are studying the nature of the electromagnetic interactions of the gold sheet with other known electrodynamically potent systems. We assume that these interactions are responsible for the difference

---

[29] Note that this turns on there being no angular difference between a partitioned trajectory, and thus no physical reason to partition the trajectory; the initial trajectory just is the final trajectory if both are the whole trajectory unpartitioned.

between process (1) and process (2). We are allowed to assume this because we treat the presence of the gold sheet as a dynamic perturbation of some original system. The electromagnetic interactions between the gold sheet and the alpha particle are the source of the dynamic change from (1) to (2) because counterfactually, we know that without it there would be no such change.[30] This allows us to say that the process that occurs between (1) and (2) is both real (deflection occurs) and has definite calculable properties (the deflection has a characteristic strength and local region). In other words, the divisions between initial dynamics, system dynamics, and our act of observing the system are provided for *by the assumed or known dynamics we use to intervene on the system.*

This generalizes. Our partition of the experimental system pre- and post-intervention is defined by that intervention (i.e., the *dynamic context* (feature (6) of §1.2). Since interventions are processes of manipulation or perturbation, and are therefore non-instantaneous, our partition is similarly non-instantaneous. This means that the boundaries between the pre- and post-intervention system are defined by a temporally extended interaction or transition. To put it more simply, because we assume that our interventions actually perturb the system, we implicitly posit that there is something going on before and after our intervention, with the intervention as (part of) the transition between the two. This is the key that allows us to apply the continuity argument.

1.4.2.4 Without Process, No Models

Our experiments have parts, including the processes of observation and intervention, and any other parts of our experiments will be defined by those processes. It remains to show that these other parts are indeed processes themselves, or at least contain processes as essential components.[31] The argument is simple: our condition of continuity applies in this case, and so those parts of experiments described by our models must be of the same ontic type as the processes of observation and intervention. Thus, they must be processes. Further, the processes described in our models will have some, but not all, of their features defined by the processes of observation and intervention used to identify and differentiate them within the experimental whole. This includes both physical and metaphysical features of those processes. E. g., electromagnetic interventions will necessitate that the model-processes are electromagnetic processes (or at least electromagnetically responsive and potent).

---

30 In fact, we could test for this to produce further counterfactuals.
31 Recall that we will later rule out that there are any other components when we move into the arguments of Chapter 2.

Although it is intuitive that all four partitioned parts of the experiment—the pre- and post-intervention system, the intervention process, and the perceptual process of observing—mereologically combine to form a single whole, we will not be making use of Part-Whole Continuity. The use of this version of the continuity condition is legitimate, but feels too easy. Instead, we will consider only Entity-Entity Continuity at each boundary between the post-intervention experimental system and the other parts. This post-intervention system is the part that we invariably will want to model in our theories, so it is the part of interest to the realist. Importantly, we have already suggested in §1.4.1 that the act of observing (the perceptual process) and the experimental system are continuous, so we will not consider this in detail.

Let us consider the boundary between the pre- and post-intervention system. Since both are defined at their partition by the intervention, both have as identifying features the features of the intervention. Namely, there is a non-instantaneous transition between the two, and both must be temporally extended and contextually defined by the type of process(es) involved in the intervention. As discussed before, we identify the evolution of the system before and after perturbing it with our intervention by noting how and in what way our perturbation changes the evolution of the system.

This means that Entity-Entity Continuity applies. The pre-intervention and post-intervention systems are only what they are in virtue of there being a particular intervention that transitions the former into the latter. Similarly, partitions between the pre-intervention system and the intervention, and the post-intervention system and the intervention, will meet the requirements of Entity-Entity Continuity.

However, there is a more striking feature of the partitions of the experimental whole that we can discuss in order to make the continuity condition utterly clear. That is, since the definition of the boundaries between experimental dynamics rests on interventions and interactions between subsystems, and since we know that no interaction or intervention is instantaneous, we can construct a mathematical representation of the system dynamics for which the boundaries are defined only within a non-zero-magnitude (though possibly arbitrarily small) region or duration. I.e., since interventions must occur and affect the system over some finite, non-zero duration epsilon, and since this defines the temporal, dynamic boundary between the pre-intervention system and the post-intervention system, this boundary is necessarily determined *only up to some finite, non-zero duration or extent delta*. Thus, (as we should suspect) not only does our condition of continuity apply, we can equally well model the partitions of the experiment mathematically such that standard mathematical proofs of continuity (delta-epsilon proofs) can be constructed. (Note this is how we discuss pre- and post-intervention systems and the partition between them. We can similarly define the partition between the system and the observer's sensory response to the system in terms of a dynamic interaction between them. Since no interaction can be instantaneous, we get a similar result: we can always model, for every experiment, the relevant pragmatic partitions in the experiment in terms of *mathematically continuous* functions defined with the temporal variable. Thus, we get mathematical continuity precisely because (and which entails) there is dynamic continuity.)

If continuity applies, then the pre- and post-intervention systems must be processually defined, just like the intervention itself. We can show this by simply noting that the applicability of the continuity condition entails that every process-feature from §1.2 (save for features (1) and (2), subjectlessness and generality) can be found in the parts of the experimental whole in virtue of two parts of this whole being processual (i.e., the intervention and the act of observation). To avoid repetition, we will only discuss the post-intervention system and the intervention.

(*Feature 3, Occurrent-Not-Continuant*): The intervention is non-instantaneous, and is defining of the resultant post-intervention system. The post-intervention system therefore is defined by its having undergone some dynamics, or by its continuing to undergo some dynamics.

(*Feature 4, Uncountability but measurability*): The intervention comes in degrees and amounts, not discrete units. The post-intervention system will therefore respond to the intervention in proportion to these degrees and amounts, and so is defined by the degree or amount of its response to the intervention.

(*Feature 5, Determinability but indeterminate*): The intervention is not statically defined by its spatiotemporal location or its causal effects, but rather has a determinable spatiotemporal extension and causal function relative to the degree of influence we think is non-negligible. The post-intervention system will therefore be determinable as "the system at the point in time at which the response to the intervention becomes non-negligible."

(*Feature 6, Dynamic contextuality*): Related to feature 5, the intervention only counts as an intervention in virtue of it having a determinable non-negligible effect on another system that responds to it. I.e., it has a necessary dynamic context. Therefore, the post-intervention system has as its dynamic context the intervention itself, plus any and all dynamics that persist through the intervention or are altered by it.

(*Feature 7, Functionality measurable in stages, phases, or changes*): the function of the intervention is its act of changing the experimental system, measurable in either a degree of change or a number of changes in a variable quantity or quality. Therefore, the post-intervention system will be functionally dependent on this degree or amount of change. It will acquire (part of) its function from how these changes are propagated forward. Notably, if the system is complex, the propagation of this functional origin in the intervention will involve many sequential functional transitions, as in the case of a nucleus bombarded with a neutron: one motion is propagated into the motions of many by sequential collisions and recollisions.

To put it simply, since our act of intervening is a process, and it is continuous with the post-intervention system, the post-intervention system must be processual in character. I.e., it is a process, with some as-yet-unknown possibility of being a process of some thing or collection of things (structures, substances, static properties, souls, and so on). We are therefore justified in calling these parts of an experiment "experimental dynamics," and we are justified in saying that any model of the experimental system must describe or identify these dynamics I.e., there is a real and inferrable process external to the observer that is described using our models.

The upshot is this: if our models hope to describe experimental systems, and so be empirically adequate, they must describe processes. This is simply because the portion of every experimental system that we have reason to believe must exist is that portion that is continuous with dynamics like our interventions and acts of observing (our sensory responses to the system or our measurement apparatuses). To quote the late physicist David Finkelstein: "In a quantum epistemology, knowledge is a record or reenactment of actions upon the system" (1996, 18).

1.5 Conclusion

The continuity argument allows us to commit to processes as real entities on the basis of experiment. The argument, which comes in three levels of complexity, enables three levels of specificity about which processes are real and legitimately inferrable on the basis of experiment and observation. These are:
- Level 1: We are allowed to commit to the existence of processes in general, because no observations are possible without *some* processes.
- Level 2: We are allowed to commit to the existence of processes in our experiments, so long as they are parts of those experiments, because the experiments are wholes with known dynamic parts.
- Level 3: We are allowed to commit specifically to the processual parts of our experiments that are (a) unobservable, and (b) described in our models. We may commit in this way provided the processes in our models are defined as the processual parts of the experiment that complete the mereological composition of the experimental whole. We may commit in this way because the completion of this whole means that the unobserved processes described in

our models are not metaphysically or physically distinguishable from the processes with which we are in direct contact.

Consider, for instance, the Bohr model of the atom, used to describe the spectral emission and absorption of light of particular frequencies by hydrogen atoms.[32] We should have no trouble admitting that the emitted light (each spectral line) is real. This is level 1 commitment: there are processes in this experiment. Level 2 commitment comes when we seek to say that there are real processual parts of our spectroscopic experiment: there is the emission process, but also the processes in the system that flow into the emission process. Level 3 commitment is most specific, and is the commitment level of interest to the realist about scientific models and theories. We state a level 3 commitment when we say that there are real processes described in the Bohr model and wholly contained in the model system to which we can commit. The transitions between energy levels are the continuation of the dynamics of absorption in the Bohr model. These transitions also continue into the dynamics of emission. Therefore, we can legitimately claim that these transitions are real processes. The Bohr orbits are not necessarily real (actually, they are impossible), but the dynamic transitions are assuredly real.

Note that what I have called the continuity condition is not actually a novel contribution, except insofar as I have named it and described it using standard Western analytic philosophy. The source of the condition is actually found in the philosophies of Taoism, Buddhism, and the Kyoto School, and in Dharmakirti in the 9th century CE. In these philosophies, a distinction that can best be characterized in terms of "external" relations and "internal" relations. External relations are just those relations we are used to in analytic philosophy: they are comparative facts or independent entities that obtain or exist independent of the nature of the relata. Internal relations, however, are those relations that mark entities that are inseparably interdefined, such that the two cannot be said to be metaphysically independent of each other. In other words, internally related entities are just those for which a defining feature of one is also partially defining of the other, i.e., entities that are continuous.

It is the work of later chapters to show how the particulars of our models follow the details laid out here. In Chapters 3, 4, and 5, I develop three such examples, making use of the arguments here and in Chapter 2 to show how historical physics conforms to the continuity argument, and is supplemented by it.

---

[32] Note that the type of atom is a historical accident, and a simplifying assumption, not a necessary defining feature of spectral experiments.

However, before this, I must first argue that things—the orthodox static, substantial opposite of processes—cannot be reasonably inferred to be real on the basis of experiment alone.

Chapter 2
Processes Underlie Processes

2.1 Introduction

In the previous chapter, I argued that we are justified in inferring the existence of processes. Importantly, this was because our experiments, and thereby the experimental systems of interest, are necessarily defined by known dynamics. We therefore infer that there must be real, describable processes in experimental systems, and that these will form at least some of the content of our models and theories. This was our positive argument for process realism.

However, I did not rule out that there is additional content in our models, or that there are non-processual parts of our experiments. While it is necessary for all parts of our experiments to at least be dynamically potent—to have the potential to undergo dynamics—it is not necessarily true that they must be actual dynamic processes.

Indeed, there is a class of arguments that there must be things—substances, structures, souls, static properties, etc.—to underlie processes. These arguments originate in Aristotle's argument that every change requires a substantial underlier, or material cause:

> Now, in all cases other than substance it is plain that there must be something underlying, namely, that which becomes. For when a thing comes to be of such a quantity or quality or in such a relation, time, or place, a subject is always presupposed, since substance alone is not predicated of another subject, but everything else of substance. [...] For we find in every case something that underlies from which proceeds that which comes to be ... (*Physics* 190a31–b9)[33]

That is, no change or dynamics can occur independent of some persistent object with which to identify the change or dynamics. I.e., no processes can exist without an underlier. I call arguments of this type "underlier arguments."

---

[33] See Cohen (1984) for a good overview of the debate surrounding Aristotle's underlier argument, and Robinson (1974) for a good benchmark discussion of Aristotle on prime matter. Note that Aristotle's argument that every change requires an underlier is one reason why many take Aristotle to be necessarily committed to the idea of prime matter, to underlie substantial change, while others argue that the underlier of substantial change need not be some further substance beyond the five elements. This very debate mirrors the analysis that I offer in this chapter.

https://doi.org/10.1515/9783110782516-007

This argument should be immediately dubious to the reader, given that some have argued that Aristotle himself considered the four elements as basic subjectless activities,[34] and more recently in view of work done by Seibt and others to show the exact opposite: we can identify and classify dynamics in the world without a subject. However, underlier arguments come in many variations, and some are especially prevalent in the philosophy of science. Namely, so-called robustness arguments.

In what follows, I argue that the essential feature of every underlier argument is the assumption, deduction, or induction of something stable within one or more experiments. This, at least, is legitimate. However, each underlier argument further treats this inference to something *stable* as an inference to something *static*. This entails that there is a static thing underlying the observed dynamics of an experiment. However, these arguments must assume that the inferred underlier is static, not merely stable. I.e., they must assume that any relative stability in the dynamics of experience is absolute stability (staticity). If they do not assume this, there can be no reason to suppose that these stable underliers are things and not processes.

Therefore, the process realist can not only refute these arguments, but can coopt them. The key is to note that the stability-entails-staticity assumption is false. In short, I argue that all we can reasonably infer is that there is something *more stable than* the experimental dynamics, rather than something static. This allows us to show that, when an inference to an underlier is warranted in the first place, it is an inference to a more stable process, not to a static thing.

Crucially, the relativity of stability is something only permissible as a feature of processes. Things are not the sort of entity that can accept degrees of stability: a thing either is or is not. It is *determinate*. However, a process is only ever determinable, identifiable by dynamic context. (cf. §1.2, features 5 and 6). Therefore, *only* a process ontology can account for the relativity of stability. This means that our negative argument against thing underliers is also a positive argument for process realism.

The chapter proceeds thusly. First, I will offer a discussion of the types of underlier arguments, a general prescription of their form, and the key differentiating factors in how they are constructed (§2.2). I also proceed through each of the underlier arguments I have collected, offering a reconstruction following my prescription of form and a refutation (§2.2.1, §2.2.2, §2.2.3). This will present us with an inductive base for constructing a general refutation of underlier arguments, namely, the rejection of the stability-entails-staticity premise (§2.3). I then conclude the chapter with a discussion of the relativity of stability (§2.4).

---

34 See Gill (1989).

2.2 Underlier Arguments and Their Types

Underlier arguments are simple in form, but are multifarious in their precise manifestation. The form is as follows:

(*Premise 1: Stability exists*): There is some stability in or related to our experience (experiments are a subset of experience).
(*Premise 2: Stability entails staticity*): Stability entails staticity, i.e., the unchanged part of an experience entails an unchanging part of that experience.
(*Conclusion: There is staticity*): Therefore, there is some static thing in or related to our experience.

Underlier arguments differ from each other in how they make this argumentative form precise and particular. To do this, they must specify:
(i) What stability in particular exists, and in what way it is stable
(ii) How this stability is not relative, i.e., how this stability entails the existence of something static, and what that static thing is

By specifying (i), underlier arguments are particularized as arguments for specific things to underlie experimental dynamics. Restricting ourselves to the domain of scientific experiments, models, and theory, this is done by appealing to three basic features of science, and physics in particular, that are suggestive of real stabilities. Namely, underlier arguments trade on (a) the stability of experimental outcomes and practices, (b) the language and models we use to describe these outcomes, and (c) that all experiments occur in a material world. Each of these features is uncontroversial when it is present, and essential to any reasonable account of scientific experiment and modeling. Thus, nearly all underlier arguments have an uncontroversial first premise, and the specifics of this first premise define what type of underlier argument is being applied. Namely, corresponding to (a), (b), and (c), we get three types:

(A) *Underliers of Experimental Practice:* Underlier arguments that trade on stability within and between experimental events and methods
(B) *Underliers of Descriptions and Model Features:* Underlier arguments that trade on stability of our language and the models we use to describe experiments
(C) *Underliers of Existence and Physical Nature:* Underlier arguments that trade on the stability of the material/physical world

However, underlier arguments that rely on these three features of science must also justify premise 2, namely how stability entails staticity (ii). In fact, each underlier argument must assume or argue that the uncontroversial stability in scientific experiments and practices entails the existence of staticity in the form of a static thing. If the argument for premise (2) is deductive, this involves the injection of additional (and often subtle) metaphysics into the argument. However, some underlier arguments appeal to inductive support for premise (2) instead. These arguments tend to fare better, and are usually the arguments that can be co-opted by the process realist. As I argue in this section, all specifications and justifications of premise (2) fail. This means that all underlier arguments fail to justify that stability is anything more than a relative feature, a comparison between different dynamics. Since processes can be relatively stable, i.e., stable with respect to other processes, this in turn means that all underlier arguments fail to rule out that the underliers of stability in our experiments are processes, not things.

2.2.1 Underliers of Experimental Practice: Stability within and between Experiments

The first stable feature of science we consider—the stability of experimental outcomes and practices—is threefold. First, in every experiment, there is persistence. In other words, every experiment exhibits some stability that persists unchanged through the dynamics of the experiment (persistence stability). Second, when similar experiments are performed many times, the outcomes of those experiments are stable. That is to say, despite the minor differences in, e.g., who by and where the experiment is performed, similar experiments produce similar results (Similar Experiment Robustness). Third, when many different experiments are performed on the "same system," certain features appear within that system as constants across all the multifarious experiments performed. These constants are said to be robust across these multifarious experiments, i.e., tolerant of the perturbations inherent in the interventions performed (Systematic Robustness).[35]

These three types of stability all appear within the experiments themselves, rather than in our descriptions of them or our assumptions about how they are manifest in the world. As such, these three types of stability support three types

---

[35] Those who employ and defend such robustness arguments are consistently opposed to the arguments I present in this chapter. For more on robustness in general, see Eronen (2015), Llyod (2010, 2015), and Schupback (2010, 2015, 2016). The robustness literature is also rich with examples from biology, climate science (of which Llyod is one), and physics, which I will not cite here. Needless to say, the common refrain of this chapter—that stability is relative, not absolute—will provide us with an interesting way of co-opting robustness arguments in favor of the process realist. Namely, since robustness is relative (it depends on a certain context and comparison), physically robust features of experiments are robust in virtue of being processes with a characteristic energy greater than the class of perturbations being considered. I return to this point in §2.3.

of underlier argument that can be categorized under a single, more-broad type: robustness arguments. Briefly, all robustness arguments trade on the fact that certain aspects of our experiments persist through the changes and dynamics that define those experiments. I.e., our interventions, and the dynamic context of our experimental setting, do not affect certain parts of our experimental system as much as they affect others. Neutrons won't respond to electromagnetic perturbations (up to a certain energy). Molecules of water do not respond to the thermodynamic perturbations involved in boiling the water, etc.

These underlier arguments then go: if there is an aspect of our experiments that is stable with respect to experimental interventions and dynamics, then it must be an unchanging thing.[36] I.e., if there is such stability/robustness within and between our experiments, there must be some static entity that underlies the experimental dynamics to explain this stability/robustness. As with all undelier arguments, it is this last inference—from stability to staticity—that is unwarranted.

2.2.1.1 Persistence Stability
2.2.1.1.1 The Characteristic Underlier Argument
Persistence stability is the sort that appears when some aspect of a system persists through experimental dynamics. Consider, for example, taking a small sample of copper sulfate and placing it within a candle flame (the intervention). The copper sulfate reacts with the other combustion components and the candle flame turns green. These changes, leading from the intervention to the change in color of the candle flame, are the experimental dynamics. However, throughout these dynamics, the candle flame's shape remains unchanged, as do the convection currents around the candle flame, (some of) the chemical products of combustion such

---

[36] There is a nuance here that is worth mentioning: robustness arguments in the literature can be both epistemic and ontic. Many arguments in the robustness literature are focused on the epistemic side: that multifarious means of measurement and description entail stronger certainty about the experimental outcomes. At first pass, these epistemic robustness arguments would seem not to make the inferences I have attributed to them. However, implicit in most of these is the ontic argument: that whatever is robust across experiments is *real.* This is because the epistemic argument is only interesting because we wish to know what entities science commits us to explicitly. I.e., our knowledge about experiments is directly linked to our ontological commitments to entities within our experiments. Thus, while many robustness arguments do not seem ontological in their first reading, nearly all of them are indeed. Those papers I have cited above make this abundantly clear, either through their allusion to ontic relevance or through their explicit claims. In what follows, I attempt to cite only those instances of robustness arguments that are seeking an ontic interpretation of experiments.

as water vapor, the chemical makeup of the combustion fuel (the Paraffin), and the production of light through incandescence. It is because of these persistent features of the system that we say that *the candle flame* changes color.

In other words, there is some entity in our experiment that persists despite the changes and dynamics within the experimental system. This entity is stable, unaffected by the interventions we perform or the goings on in the system. Moreover, there must be such a persistent entity in every experiment. While all experiments involve some change in the system-plus-observer continuum, this change is necessarily contrasted with some aspect of this continuum that remains unaffected by the interventions performed.[37] It is a natural enough assumption to make that this persistent entity is a static entity, i.e., a thing. Therefore, the thing realist argues that things underlie experimental dynamics because every experiment must include a persistent thing. In more precise form, this argument goes:

(Persistence Underlier Argument):
($P1_{PS}$): There must be a persistent entity within (an/every) experiment.
($P2_{PS}$): A persistent entity is necessarily, or necessarily contains, a static entity (stability entails staticity).
($P3_{PS}$): A static entity is a thing (there are no static dynamics—I omit this premise in most cases).
(C_{PS}): Therefore, there is at least one thing in (an/every) experiment.

($P1_{PS}$) is uncontroversial, as is ($P3_{PS}$). For this argument to work, we need only to justify ($P2_{PS}$), that stability entails staticity.

To do this, thing realists have offered many justifications in the past. For example, one argument common to early modern philosophers like Descartes and Locke goes that a persistent entity consists of properties that can change and properties that cannot.[38] It is because there are such unchanging properties that an entity is

---

[37] Oddly enough, this is a key part of Wesley Salmon's so-called causal process account of causation: the recognition of a "mark" that is transmitted in every causal process (W. Salmon 1984, 1994; cf. Reichenbach 1935, 1956). Phil Dowe (1992, 1995, 2000, 2003) also adopts this in his slightly different account of causation, where "marks" are given physical definition as "conserved quantities." Arguably, neither account can be considered a pure process-realist account, since neither account takes seriously that processes are basic. However, both have value for informing the position of the process realist on causation.

[38] See Descartes (*Meditations* I) and Locke ("On Identity and Diversity") for their relevant views. For a more detailed gloss on Descartes' view in the context of debates on perdurance and endurance, see Gorham (2010). For a more detailed gloss on Locke, see Strawson's (2011) commentary in the cited work by Locke above. Locke's view is couched in terms of personal identity, not strict haeccity or the nature of physical entities in general, but it is no less relevant for this.

capable of persisting. The unchanging properties of the entity—its static essence—explains its persistence.[39] Put another way, staticity grounds the stability of persistence, and so persistence entails the existence of (some) staticity. In the case of Descartes at least, this argument was meant to provide a foundation for scientific theory and practice (Moore 2012, ch. 1).[40]

More recently, the literatures on temporal parts and personal identity contain implicit justifications of premise 2 of the persistence underlier argument. That is to say, because these literatures often rely on persistence arguments to justify their core points, these literatures contain implicit appeals to various methods of justifying premise 2.[41] For example, one account of personal identity goes that a person is identified by those parts that remain unchanged through mental and physical changes.[42] Similarly to Aristotle's argument, as long as there is some part of the person's psychology that remains unaffected by changes in their mental or physical state, they can remain the same person through these changes.[43] Another view holds that personal identity is maintained by there being a single biological organism that remains unchanged across time.[44] Yet another view holds that a single

---

[39] This argument is present both in Scholastic and contemporary philosophies, particularly in the Scholastic interpretations of Aristotle and up to more recent debates about the nature of temporal parts and persistence. As already mentioned, Aristotle (190a31–b9) is the first instance of an argument that every change is grounded in a material substrate. More recent debates about persistence are found within the literature on perdurance and endurance, largely as a result of first the seminal work by McTaggart (1908, 1927) and by later work by Lewis (1976, 1986, especially p. 202). Contemporary authors such as Most interestingly for our current discussion, Wasserman (2016) offers an argument that the debate about temporal parts is the result of a conflation of the ontological question (whether objects have temporal parts) and the epistemological question of whether objects persist in virtue of those temporal parts. In other words, Wasserman explicitly shows that the debate about temporal parts is motivated by the desire to explain persistence in terms of static states and parts of being. See also Wasserman (2003, 2004, 2005, 2006). The most explicit recognition of the persistence argument comes in Wiggins (1980, 2001).
[40] Moore also remarks that Descartes's very epistemological project is to discover that which is "stable and likely to last" (2012, 29). This suggests the further point that, for early modern philosophers following in the Cartesian tradition, one of the core projects for grounding scientific theory was to provide a vindicating epistemology to show how science produces stable claims, in addition to being about stable entities and facts.
[41] C.f. Wasserman (2016), who argues that in the literature on persistence, there is an implicit conflation of the explanation of persistence and whether or not entities have temporal parts.
[42] See, for instance, Swinburne (1984, 21).
[43] For more on this see Garrett (1998), Hudson (2001, 2007), Johnston (1987b, 2016), Lewis (1976), Nagel (1986, 40), Noonan (2003, 2011), Parfit (1971, 1995, 2012), Perry (1972), Shoemaker (2008, 2011), and Unger (1979; 1990, ch. 5; 2000).
[44] For more on this "brute physical identity" view, see Ayers (1990, 278–292), Carter (1989), Mackie (1999), Olson (1997), van Inwagen (1990, 1997), and Williams (1956–57, 1970).

person is identified as the recurrent referent for the protagonist in a story or narrative.[45] In all of these, diachronic persistence is explained by there being something—psychological core, brute biological fact, or narrative referent—that remains unchanged through the inevitable dynamics of progression through time.

Similarly, the literature on temporal parts, especially those papers focusing on the so-called problem of change—alternatively and tellingly renamed the "problem of temporary intrinsics" (Lewis 1986a, 203–204)—contains reference to static underliers of persistence. Endurantists hold that change is defined as a difference of state between two times. The change in a system, then, is grounded in there being some constant between those two states: the endurant object. E.g., a bud changes into a flower over time, and we can say this because throughout the change from bud-state to flower-state, other aspects of the plant organism remain identical (only the shape changes).[46] Of particular interest amongst endurantists is Melia (2000), who argues explicitly in favor of enduring things over processes. In contrast (if we can call it a contrast at all), the perdurantist and stage theorist argue that it is only because objects have temporal parts that they can be said to change. E.g., the plant is said to blossom because there is a temporal part of the plant that is a bud, and a temporal part that is a flower, and these temporal parts appear within the same four-dimensional object.[47] Interestingly enough, endurantists criticize this position because they see it as not describing real change.[48] Regardless of

---

45 For more on the narrativist view, see Schechtman (1996, 2001) and Schroer and Schroer (2014). Strawson (2008) and Olson and Witt (2019) criticize the narrativist view. See also DeGrazia (2005).
46 For more on Endurantism, see Fiocco (2010), Merricks (1994), Hinchliff (1996), Zimmerman (1998); see also Johnston (1987b), Haslanger (1989a, b), and Lowe (1987), for adverbialist positions (derived in part from Sellars' 1952 work).
47 For more on Perdurantism, see Hudson (2001), Lewis (1971, 1976, 1986), Quine (1950, 1960), and Robinson (1982). Stage theory offers a slightly different explanation of change, although one with a similar spirit. For stage theory see Sider (1996, 2000, 2001) and Hawley (2001, ch. 1). For alternatives to endurantism and perdurantism, see Brower (2010), Ehring (1997, 2001), MacBride (2001), Seibt (1997, 2008), and Simons (2000). Note that in some of these alternatives, the difference between the account and the typical forms of perdurantism and endurantism is somewhat minimal, as in Simons (2000) which is endurantist but about Whiteheadian universals. Most interesting for our discussion here, Klein (1999) argues that genuine change cannot be captured by either endurantism or perdurantism. Klein is therefore most in line with the process-realist position, although not an explicit advocate of it. See also Meincke (2018a, b).
48 For more on this criticism, see McCall and Lowe (2009), McTaggart (1927), Mellor (1998, section 8.4), Oderberg (2004), and Simons (1987). The argument offered in various forms by these authors is quite similar to arguments offered by, e.g., Henri Bergson (1994 [1896]) and William James (1981 [1890], 1977 [1909]) that continuous experience cannot be reconstructed from static stages or states put in order. A similar argument is offered by Aristotle (anachronistically) and his interpreters regarding the real numbers, namely that no number of points can be used to reconstruct the contin-

these differences, in each explanation, the persistence of a manifestly changing object is explained by there being something unchanging—the core 3D object, or the 4D object—that either gains/loses new properties or else has static parts at different times that manifest different properties.[49]

The key to these defenses of premise 2 is that they are all deductive. Persistence entails staticity because of a priori principles about what sort of entity can (in the minds of the thing realist) persist. That is, (a) only objects with property-sharing temporal parts can persist, (b) only objects without intrinsic temporally-indexed properties can persist, (c) only objects with some unifying and unchanging biological referent can persist, and so on. Refuting these arguments, then, is as simple as rejecting the premise of the discussion.

2.2.1.1.2 The Refutation of Persistence Underlier Arguments

The persistence underlier arguments rests on establishing deductively that persistence entails something static. However, this is false. Persistence does not entail the existence of a static unchanging entity. Whatever persists in our experiments is, admittedly, unchanged by our interventions or the dynamics within the system. However, this does not entail that these persistent entities are *unchanging*. In other words, persistence within an experimental system alone is insufficient to conclude that there is something static, not when we could equally conclude that there is a persistent set of dynamics that are unaffected by the interventions we perform.

Therefore, to justify (P2$_{ps}$), the thing realist must adopt additional deductively stronger premises about the nature of persistence. We saw a few examples of this

---

uum. Similarly, the authors cited here argue that temporal parts cannot simply be arranged in order to reconstruct the continuous flow of change. Of these, Orderberg (2004) is most in line with the process realist, in that he argues that there is no puzzle to be solved. Heller (1992) argues that there is such a puzzle (the problem of change), and that there is genuine change to be found in endurantism. See Lombard (1994) for a response to Heller. See also Botterell (2004).

49 I confess, I long found this debate troubling, as have many process theorists. Change is not to be defined by reference to static things. Change is change. The flower doesn't grow from the bud because bud and flower are different stages of one entity, the flower grows from the bud because the bud blossoms. The change is primitive, observable, and inalienable. The reason why there is a "problem of change" in the first place is precisely because thing realists have long dominated philosophical and scientific discourse. For good arguments to this effect, see Seibt (1990), who argues that the problem of change is the result of a certain number of a complex of 22 assumptions made in substance (or thing) metaphysics. For more recent arguments against the problem of change, see Hansson (2007), Hofweber (2009), Rychter (2009), and Seibt (2008). Raven (2011) offers a response, arguing that there is indeed a problem of change. See Einheuser (2012) for a summary of the debate.

above. If such premises were adopted solely for the sake of establishing that persistence entails staticity, the premises would beg the question. For example, consider the position of the radical endurantist who holds that persistence can only be interpreted as the duration of an unchanging structure or collection of unchanging properties. This assumption in this context presumes the conclusion: that persistence entails the existence of something static. Similarly, if one argues as the early modern philosophers did that persistence results from the existence of a core or primary set of properties that are unchanging, one similarly begs the question.

The situation is slightly complicated, however. Most of the authors I surveyed above are interested in resolving an existing problem within thing/substance ontology. In short, the assumptions about persistence are not merely meant to establish the existence of static things, but also are used to meet various other explanatory needs.

For example, it seems natural to suppose that my ability to reidentify some persisting portion of an experimental system is grounded in something like mathematical identity. In other words, I can identify this spectroscope now with that spectroscope then because the respective systems each have some set of properties X and Y for which equation $X = Y$ holds true. This would be one more nuanced reason to suppose that the persistent system is defined by some collection of properties that are unchanged.

There are two problems with this particular account of persistence. First, even if we must understand various instances of persistence in this way, this does not entail staticity of the cross-temporally identical properties. It could very well be the case that $X = Y$ is true across the relevant duration, but ceases to be true across greater durations. I.e., the property X could be time-scale sensitive, in that for durations less than some value T, X is unchanging, but for durations greater than T, X changes. For example, we might say that a radioactive atom is like this: persistent (with moment-to-moment identity of atomic number) for time scales less than the average half life of the element, not persistent for greater time scales.

Second, and more interestingly, it seems unlikely that such a static definition of persistence will ever successfully track the physics of any experimental system, even the persistent aspects of it. This is to say, the history of physics seems to rule out that any system is absolutely persistent. In virtue of being characterized by energy, physical systems are susceptible to energy fluctuations. We may discuss, e.g., the motions of molecules in a gas when we heat or cool the gas. We may even say that the molecule is a persistent entity in this gas. However, we know that the molecule's structure and composition are only persistent and stable in the gas because we are not perturbing the system with enough energy to break this structure. Our claim that the molecule persists is true only because we are observing and experimenting with its structure in a way that does not destroy it.

All systems in physics are perturbable in this way. Even properties like charge in an electron system can be gained or lost under the appropriate perturbations (e.g., the introduction of appropriate fluctuations in the quantum electrodynamic field). While there are some systems that we cannot manipulate directly, we nevertheless know that their persistent aspects are manipulable and perturbable.

We are therefore put in the position of asking what benefit could be gained from presuming that the persistent parts of our experiments are static. If we know that they are only unchanged by the experimental dynamics, why should we suppose that they are *unchanging*? There are, perhaps, good reasons to suppose this in metaphysics. An unchanging persistent provides a clear and effective means of characterizing cross-temporal reidentification. An unchanging persistent allows for the definition of persistence in terms of definite determinants like static properties of states at a moment in time. However, when we move to the domain of physically realized experiments, we need to admit that all of the persistent entities we recognize in those experiments are not necessarily unchanging persistent entities. There may be some such, but the burden of proof lies on the thing realist to produce examples.

Moreover, process ontology can admit of cross-temporal reidentification such that processes can successfully ground statements of persistence. The fact that one possible dynamic shape (C.f. §1.2) is the cyclic process proves this. Cyclic processes like pendular or harmonic motions can allow us to mark the coincidences between the cyclic process and another process. Once we have marked these coincidences, we can further mark out various features of the non-cyclic process. If that process has the same processual features between coincidences, then we can say it is the same process. Finally, once we have cross-temporal reidentification of a process, we can use this process as a comparison between processes. This allows us to say that one process is "faster" than the other, or is otherwise less stable. The more stable process can then be said to persist through the less stable process.

Indeed, since most apparatuses built to measure or record time, and so to record persistence at a basic level, involve cyclic processes should suggest to us that persistence is not necessarily a thing-specific concept. Everything from light clocks to grandfather clocks make use of cyclic or at least recurrent dynamics in order to measure the duration of events. It would therefore be unwise to suppose that the persistence underlier argument can successfully rule out that the persistent parts of our experiments are processes rather than things.

Finally, as I argue later, there is reason to suppose that claims about stable persistence *should* be understood in terms of processes and not things, not merely that we could do so. This is because all stability statements involve a relativity assumption: something is stable with respect to something else. In the case of persistence, we can reconstruct this relatively simply: All persistence claims are true in virtue of the persisting entity being compared to some smaller, cyclic process. E.g., I would say that a table persists in virtue of having comparably unstable systems that undergo noticeable change much faster, systems like light clocks, decaying atoms, candle flames, and air currents. This means that for every persistence claim, we can interpret it in terms of relative energy or time scales of comparable systems. A table typically lasts for years in roughly the same structure. A candle flame changes in seconds, and candles change in hours. Persistence, then, is relative to time scales. Process ontology can account for this; processes are assumed to be temporally extended with characteristic time scales. Thing/substance ontology will find it difficult. While a substance ontologist may allow for the generation and destruction of substances, the relativity of the time scales of such creation and destruction events comes only once we understand such events as temporally extended, and only in the context of specific interactions. Thus, the persistence underlier argument either fails (because it can't rule out that processes are persistent underliers) or can be co-opted (because we can alter it to infer the existence and features of further processes or processual features).

2.2.1.2 Replication Robustness
2.2.1.2.1 The Characteristic Underlier Argument

The second type of stability we see in our experiments—single-experiment robustness—is a feature of the repetition of an experiment. Namely, when we perform one intervention on similar systems, we notice that the observed outcomes are similar. For example, suppose that you, I, and another scientist all take a candle, light it, and introduce a sample of copper sulfate into the combustion region using a pair of tweezers. All three of us will observe that the candle flame, once yellow, now burns green. Moreover, all three of us will observe that our particular flame's shape, its incandescence, the water vapor rising from it, etc. remain unchanged. In short, all three of us observe similar experimental outcomes, both in terms of what changes in the system and what remains unaffected as a result of our similar interventions. I.e., our experimental outcomes are stable in replication.

According to the thing realist this similarity between experimental outcomes must be explained in terms of some feature of the replicated experiment that is statically similar in all three instances. For example, in our candle flame experiment, we must look for some aspect of our three experiments that identifies the systems as "the same" system. Similarly, we must look for some aspect of the experimental outcomes with which to identify each of our outcomes as "the same" experimental result. The thing realist argues that these identifying features must be static things present in the system because we may only consistently refer to such static things. Thus, the stability of replicated experiments (their outcomes and features) entails the existence of a static thing, and so static things must underlie experiments. Put more precisely, this argument goes:

(Replicability/Similarity Underlier Argument):
($P1_R$): Identifiably similar or replicable experiments produce similar results.
($P2_R$): Experiments are identifiably similar because they share static things within their systems to which experimental descriptions refer and on which experimental results depend (stability entails staticity).
(C_R): Therefore, there are static things that underlie similar or replicable experiments.

Once again, ($P1_R$) is uncontroversial (when it obtains) and it is only ($P2_R$) that requires support.

Notice that there is far more work going on in ($P2_R$) than it might initially seem. Support for this premise must include a defense of the claim that experiments are verifiably similar *as the result of* some feature of the system and the set-up of the experiment. Such a defense can be found in the literature on replicability.[50] However, we must note that this literature often relies on arguments from other domains to support claims about what makes two experiments similar, or one experiment replicable.[51] An important example comes from Schmidt (2009), who in turn draws from Hendrick (1991) and Radder (1996, 2003, 2006, cf. 2009, cf. 2012). Both define one of the key functions of exact replication (i.e., the replication I have described here) as the verification that previous experiments were not the result of chance or specific laboratory conditions. Exact (direct, concrete, literal) replication is contrasted with inexact or conceptual or constructive replication in which variables in the experiment are purposefully altered in order to fur-

---

[50] See especially Nosek, Spies, and Motyl (2012) for summary and paradigmatic accounts. Steinle (2016) criticizes their argument that science must be replicable from an historical perspective.
[51] For example, Gómez, Juristo, and Vega's (2010) account of replication defines five manners in which two experiments may be identical (and thus five ways for replications to diverge from their parent experiment): spatiotemporal location, experimenter, apparatus or interaction between experimenter and system, measurement conventions and "operationalizations," and so-called population properties. They draw these types from a survey of literatures, including existing examples of replicated studies, and so their account is reasonably comprehensive of the literature.

ther build upon previous experiments.⁵² What is described as inexact replication will be subject of §2.2.1.3.

2.2.1.2.2 The Refutation of Robustness/Similarity Underlier Arguments

Much like the persistence argument before it, the success of the replicability/similarity argument rests squarely on the success of ($P2_R$), the stability entails staticity premise. In much the same way as the persistence argument, the process realist can refute or co-opt this argument by rejecting or modifying this premise. The modification is this: experiments are plausibly similar not because of their shared static things, but rather because of their shared dynamics. When we compare two experimental systems, we appeal primarily to the specifics of the way we intervene on the system, what sorts of dynamic controls we have in place to prevent outside interference (perturbations from an uncontrolled source), etc. Our appeal to the similarity of two experiments is also the result of noticing that we actually observe the system in the same way: when we observe similar outcomes, we are interacting with our individual experimental systems in a similar way to each other.⁵³ In our candle flame example, we all interact with our particular flame electromagnetically, and this interaction has a characteristic energy, rendering our vision of the flame "green." In short, we appeal to the dynamic features of the experiments primarily.

Moreover, supposed non-dynamic similarities between experiments can also be defined in terms of the processes within the system. For this, we need only recall that persistent features of the system can be understood in processual terms. If persistent features of the experiment—the types of tools used, the variables modeled, etc.—can be processually defined, and the experimental dynamics are processes, there is no need to suppose the existence of things in order to explain similarity or replication. Thus, piggybacking on the response to the persistence

---

52 For more on the distinction between exact and inexact replication (and the benefits of each), see Keppel (1982), Lykken (1968), Sargent (1981), Schmidt (2009), drawing from Gómez, Juristo, and Vegas (2010) and Radder (1996, 2003, 2006, 2009, 2012).

53 This account is similar to that found in Norton (2015), who argues that there is no universal principle of replicability, but rather that replicability is established by similarity in background facts of experiments, things like the interventions performed and our pre-existing knowledge of how these interventions affect various features of experimental systems. While not explicitly process-realist, I believe this account and argument can be easily modified to be so by simply noting that the background facts to which Norton appeals will inevitably be processual in character. Processes have the uniqueness and contextuality of identification necessary to provide for Norton's local-inference account of replicability because they are general entities defined in part by a dynamic context.

underlier argument, the replicability/similarity underlier argument is easily refuted.

We might have expected this, taking the continuity argument of Chapter 1 seriously. Experiments are basically and most modestly dynamic continua: temporally extended dynamic activities in the world beginning with us intervening and ending with us observing the evolution of a system. We might further posit that there are stable features of the experiment that allow us to replicate the experiment. However, fundamentally, we cannot replicate an experiment by reproducing these stable features alone. We only replicate an experiment successfully when we perform similar interventions and engage in similar acts of observing a system.

Indeed, the similarity between experiments is relative to these dynamics. Suppose I (1) observe one candle flame with my eye, and (2) observe another with an infrared sensor, leaving the interventions the same in both instances. The similarities between (1) and (2) are apparently contextually dependent on exactly this difference in dynamics: the act of observing the system in both (1) and (2) is similar in that both involve electromagnetic interactions, but different in what subclass of electromagnetic interactions are being considered. If we wished to say that we had replicated an experiment revealing that candles emit light across a specific spectrum, both (1) and (2) count as relevant experiments for this. However, if we wished to say that candles can produce thermodynamic fluctuations in the air, (1) is less relevant than (2) *because of the difference in the nature of our dynamic observations*.

Therefore, as before, the replicability/similarity argument for thing underliers fails to rule out that the similarity between two experiments is a similarity of dynamics, not statics. Moreover, once more, we have reason to suppose that a process account of this similarity will prove similarity because of relativity considerations. Namely, because similarity between experiments is relative to the context in which we interpret them (and use them to draw inferences),[54] we have reason to suspect that there can be no absolute similarity or difference between experiments. Process ontology admits of such context dependence as a defining feature (§1.2).[55]

---

[54] This is a greatly discussed, and sufficiently proven point in philosophy of science. Namely, many authors have remarked on the context-sensitivity of the interpretation and usefulness of experiments. See especially Rehg (2009a, b, c), who both surveys the literature and presents an argument to the effect that the context of practice and purpose—i.e., the dynamics of intervention and observation techniques—are key components to defining similarities between experiments (to allow for the communication between experimentalist groups and/or theorists). See also Davidson (2001).

[55] While clearly not committed to process realism, Ross (forthcoming) offers compelling arguments (cf. especially 8–9) to the effect that we are actually better off in causally complex systems

Thing/substance ontology will need a good deal of additional metaphysical structure in order to recover this context dependence.[56] Once more, an underlier argument can be easily co-opted into an argument for processes.

2.2.1.3 Systematic Robustness
2.2.1.3.1 The Characteristic Underlier Argument
The third form of stability we observe within our experiments is the stability of features of a single system under many different interventions.[57] When we perform many different interventions on a single system, we inevitably observe many different dynamics within the system leading to many different observed outcomes. However, regardless of these differences, the systems upon which we perform these interventions remain comparable. I.e., each of the systems retain certain features that are unchanged by each of our many interventions, and the corresponding dynamics within the system. In short, the system has features that are robust in our experiments.[58]

For example, suppose I introduce copper sulfate in my candle flame, while another experimenter introduces strontium chloride, and another introduces calcium chloride, and another introduces potassium sulfate and potassium nitrate (in a 3:1 ratio). In each of our experiments, we will observe a different colored flame: green, red, blue, and violet respectively. However, in each of our experiments certain features of our experiments remain unchanged. Namely, the shape of the candle flame, the fact that combustion is occurring, the fact of incandescence, the existence of certain combustion features, etc.

This is the simplest form of systematic robustness. In each of our different experiments, we arrive at different experimental outcomes while certain aspects of

---

describing the key features of the experiment in terms of a single causal process (a pathway or mechanism) for the sake of replicating and understanding the experiment as replicable or its results as communicable.

56 Though not explicitly process-realist, this is a central point in recent work on the nature of data, e.g., in Bokulich (forthcoming), and Bokulich and Parker (2021). The key for these approaches to data is to recognize that there is an interaction between data and models, such that it is impossible to describe data as if it is fixed with definite properties. As a result, Bokulich and Parker argue that data has a phylogeny, or an evolution through history as it is progressively changed by ongoing research and modeling practices. In short, it is precisely because data is context dependent that we cannot understand it as a fixed thing, and must instead move toward understanding it as dynamic in character.
57 As already mentioned, this has been called inexact, constructive, and conceptual robustness or replicability of an experiment.
58 This is perhaps the strongest argument in favor of thing underliers, apart from the unification argument to come.

the system remain unchanged. These unchanged aspects are called the robust features of the system. The corresponding underlier argument that trades on this robustness can therefore be thought of as an induction over the first underlier argument we considered. Namely, we have four different experiments, and therefore four different potential persistence arguments. We then see that in each of these experiments, the same entities are persistent, and so the persistence arguments all share the same particular claims about what persists. By induction, we therefore suggest that these entities must be the persistent entities. Crucially, we do not claim that they are persistent on the basis of purely metaphysical reasoning, e.g., that these entities must be present in any persistent system. Rather, we support that these entities must be the persistent entities in the system by induction over many experiments. We therefore arrive at a modified version of our earlier persistence argument:[59]

(Systemic Robustness v1):
($P1_{SR1}$): There must be a persistent entity within (an/every) experiment.
($P2_{SR1}$): In many interventions on the same system, the same persistent entity (entities) appears (contingent experimental fact).
($C1_{SR1}$): Therefore, this persistent entity must be unchanged by all experimental dynamics involved in each of the experiments, e.g., it is unperturbed by any of the many interventions (inductive base).
($C2_{SR1}$): Therefore, this persistent entity must be unchanging, because it is unchanged by many dynamics and interactions (inductive generalization).[60]
(C_{SR1}): Therefore, there is a static thing in (an/every) experiment.

In short, we have replaced ($P2_{PS1}$) (persistence stability entails staticity) in the persistence underlier argument with an induction over many experimental interventions. Namely, we induce that since the same entity remains unchanged under several interventions and experimental dynamics, that entity will be unchanged under all dynamics in the system. I.e., the stability under many dynamic changes

---

[59] A good historical example of this sort of argument is the investigation of the rate of downward acceleration of material bodies. The candle flame example I provide is simpler, and is historical as well, but is perhaps a little simplistic.
[60] It is here that I should reiterate that this particular underlier argument is found primarily in the philosophy of science, not in the literature on metaphysics itself. As such, the discussion here is targeted at philosophers of science who discuss robustness as a means of achieving scientific realism.

exhibited by this entity entails its staticity.⁶¹ This induction is therefore no more than an inductive means of supporting premise 2 in the persistence argument.

The success of this argument will rest on the strength of the inductive step. ($C2_{SR1}$) can only work if ($C1_{SR1}$) is assured. Unfortunately, our simple example of robustness—that of many different alterations of the color of the candle flame—does not produce a sound induction. This is because our many interventions on the system all have roughly the same character: we introduce a new element into the combustion reaction and observe the change in color of the candle flame. As a result, the similarities between our experiments are manifold; we could conclude that anything from the shape of the candle flame to the chemical nature of the combustion products is a persistent and unchanging entity within the system. In many ways, while we are performing manifestly different interventions on the candle flame system, this experiment (and the resultant robustness v1 argument) is too similar to the replication-robustness argument of the previous section.

Therefore, we might improve the robustness argument by making the differences between our experiments more striking. For example, rather than four different experiments that all manipulate the same dynamics within the system— e.g., the production of light of a specific color in the candle flame—we might instead manipulate many different aspects of the system. Suppose that I manipulate the color of the candle flame, but another uses a different sort of wax for their candle, and another uses a differently shaped candle stick. In each of these cases, we disturb different features of the candle flame system: I disturb the

---

[61] The key difference between the persistence argument and this robustness argument lies in the induction performed in the latter. The persistence argument and its critical premise—premise 2 that stability entails staticity—were supported deductively from first principles. This meant that the proponent of the persistence argument needed to support their second premise with quite powerful (and dubious) assumptions. However, the robustness argument seeks instead to support its critical premise through induction. This means it does run the risk of begging the question, but does fall prey to the problems of induction. Notable amongst these problems is the problem of underdetermination, which is a problem unique to robustness arguments in particular. See Duhem (1954 [1914]), Goodman (1955), Mill (1900 [1867]), and Quine (1975, 1990) for the typical locus of underdetermination. See Laudan (1990) for a criticism of the radical scope of underdetermination arguments. See also Belot (2015), Norton (2008), Stanford (2001, 2006) for less radical approaches to underdetermination (though each accepts the underdetermination problem, none see it as requiring radical skepticism about the scope and rationality of science). See Chakravartty (2008) for a defense of realism against underdetermination. Note that the problem of underdetermination is not one I am seeking to resolve in this discussion, but I do believe process realism presents a novel solution in virtue of being independent from the thing-realist's robustness arguments. In other words, processes are not underdetermined because we do not need to inductively infer their features in the same manner that we infer the features of things.

color, while the other experimenters disturb the ratios of the chemical products of combustion and the shape of the flame respectively.

Our robustness argument now looks like:

(Systemic Robustness v2):
($P1_{SR2}$): There must be a persistent entity within (an/every) experiment.
($P2_{SR2}$): In many interventions on *different features/aspects* of the same system, the same persistent entity appears (contingent experimental fact).
($C1_{SR2}$): Therefore, this persistent entity must be unchanged under all experimental dynamics involved in each of the experiments, e.g., it is unperturbed by any of the many interventions (inductive base).
($C2_{SR2}$): Therefore, this persistent entity must be unchanging under all experimental dynamics because it is unchanged under many *dissimilar* dynamics of a single system (inductive generalization).
(C_{SR2}): Therefore, there is a static thing in (an/every) experiment.

The strength of this argument (compared to version 1) comes from the strengthening of the inductive step. In short, the fact that the stability we observe is unchanged in multiple dissimilar interventions entails that it is unchanged in dissimilar dynamics. This in turn suggests inductively that, no matter what is going on in the system dynamically, that stable entity will be unchanged. Were we capable of producing a list of all possible dynamics the system could undergo, this argument would provide a sufficiently strong formalism for determining if that system were truly static. In essence, version 1 of the robustness argument performed an induction over only a single case, while version 2 performs an induction over many cases.

However, this second argument is still not as strong as we might like. The stability in the system may just be a quirk of the system itself. Since all interventions are performed on this same system, we have no way of knowing if the observed stability will persist outside of the dynamics of that system. This means that ($C2_{SR2}$) is dubious; even if we could say that our observed stability is unchanged under all interventions and dynamics within our single system, this does not thereby entail that it is unchanged under all interventions and dynamics. The inductive generalization ($C2_{SR2}$) requires the universality of the inductive base: the unchanged entity must be unchanged in every system under any intervention.

We therefore make one final improvement to the Systemic Robustness Argument. Suppose that we not only make the differences in our interventions on a single system more striking, but we also perform these manifold interventions on different systems entirely. For example, I light a candle, and another experimenter ignites wood, and another ignites gasoline. No longer are the components of our experimental systems the same, nor our interventions. If, in such a case, we all observe the same stabilities, this would provide strong evidence that this stability is present universally—it persists both through different interventions and across different physical systems. In our example, we might notice that in all cases, oxygen is always present in the combustion of the system. Thus, we might conclude that oxygen is a static thing that underlies the process of combustion.

This gives us the final form of the Systemic Robustness Argument:

(Systemic Robustness Argument):
($P1_{SR}$):   There must be a persistent entity within (an/every) every experiment.
($P2_{SR}$):   In many interventions on different features of one system, the same persistent entity appears (contingent experimental fact 1).
($C1_{SR}$):   Therefore, this persistent entity must be unchanged under all experimental dynamics involved in each of the experiments; it is unperturbed by any of the many interventions on the single system (inductive base).
($C2_{SR}$):   Therefore, this persistent entity must be unchanging under all experimental dynamics because it is unchanged under many *dissimilar* dynamics of a single system (inductive generalization).
($P3_{SR}$):   In interventions and experiments on different systems, the same persistent entity appears (contingent experimental fact 2).
($C3_{SR}$):   Therefore, this persistent entity must be unchanged under all dynamics involved in the separate systems (inductive base).
($C4_{SR}$):   Therefore, this persistent entity must be unchanging under all dynamics in all systems.
(C_{SR}):   Therefore, by ($C2_{SR}$) and ($C4_{SR}$), there is a static thing.

This version of the argument involves two inductions. The first induction is meant to establish that there is an entity that is unchanging within a single system. This is just what was done in version 2. The second induction is meant to establish that this unchanging thing is unchanging outside the context of that single system. This second induction is not performed over specific interventions, but rather over systems. The inductive step, then, is to infer that if the persistent entity is unchanging within many systems, and it is unchanging under many interventions on each of those systems, it is unchanging universally. A universally unchanging entity is nothing more than a static entity, a thing. Thus, by combining these two inductions, the thing realist produces a strong argument for the claim that there is at least one static thing to underlie experimental dynamics.

2.2.1.3.2 The Refutation

Crucially, the robustness argument is inductive. Refuting it, therefore, is not as simple as noting the failure of its deductive principles. While this does means that the Systemic Robustness Argument inherits various problems of induction, it is much stronger as an argument for static things than the other arguments we have so far considered. However, a similar line of reasoning is still available to the process realist. Namely, the Systemic Robustness Argument still fails to rule out that the persistent underliers are processes and not things. This is because the second induction—induction over all experimental systems—is illicit, even if the first—induction over interventions and perturbations—is justified. It will prove difficult to provide the necessary general principles with which to reasonably model all possible experimental systems. Without such principles, we cannot know that we have successfully ruled out experimental systems in which the persistent entity we are considering fails to persist. In short, the second induction will be underdetermined.

The first induction—that a stable entity persistent through some dynamics in a system will persist through all dynamics in a system—is somewhat justified. If we define our "system" by some characteristic set of dynamics, it becomes very easy to exhaustively test those dynamics, or at least those types of dynamics. For example, in our candle flame example, we might say that a candle system is characterized by combustion, convection, capillary action, incandescence, evaporation, melting, and cooling processes. Even if the token instances of these processes in a particular candle system differ subtly from those in another candle system, they will still be instances of the same dynamics. Thus, if we test each of these dynamics and observe the same emergent stabilities, we can reasonably say that these stabilities are stable with respect to *all* of the dynamics within the candle system.

The reason we need the second induction, then, should be apparent. Namely, an entity that is stable with respect to the dynamics in a candle system will not thereby be stable with respect to all dynamics. For this to be true, there would need to be no dynamics in the world—no interactions, motions, etc.—beyond those in the candle system. By testing for these same stabilities in systems other than the candle, we widen the scope of the dynamics with respect to which the considered entity is stable. Thus, we build evidence that this entity is not only stable, but static.

However, unlike the finite collection of characteristic type-dynamics within a candle system, the dynamics within the world are infinite. Any change, any interaction, any motion is a type of dynamics within the world. As a result, our induction from the stability of the considered entity to its staticity will necessarily be an induction from finite facts to infinite scope. This means that, in order to perform the second induction necessary to complete the robustness argument, we would need some sort of universal material fact[62] or principle to justify that the dynamics we have tested are not meaningfully different from those we haven't tested.

No such material fact is available. Every system we have observed to date contains stabilities that we know could be destabilized. The only thing preventing us from performing these destabilizations (when we are, in fact, prevented from this) is the sheer amount of energy required to perturb the relevant systems. However, this is merely a practical limitation. In principle, there is always an energy threshold past which some stability in the world can be destabilized. The thing realist, accordingly, would need to argue here that there is an absolute energy threshold in the world, past which no system's energy could ever increase. In addition, even if such an absolute threshold were to exist, one would still need to describe an entity that is stable at (exactly!) that energy. When such an argument is made, I will accept the existence of at least one static entity. However, contingent upon us being so far incapable of stating such universal facts about absolute stability, the second induction fails: we cannot say that any stability in the world is absolute (i.e., static).[63] If there is no such absolute stability, we can once more turn to our refutation of the Persistence Underlier Argument to argue that the persistent underliers of our experiments might very well be processes and not things.

In short, the robustness argument fails to establish that there are static things without serious and (as yet) unprovable assumptions about physical contingencies. I.e., once again, we cannot justify that stability entails staticity. Moreover, this failure accords with an important point about our experimental practice. Namely, that stabilities in an experiment, i.e., each process in an experiment, has a characteristic energy threshold. I.e., the energy of the probes we use in our experiments determine what we treat as dynamic vs. stable parts of a system. This point is critical for the epistemological aspects of pure process realism: how we determine what sorts of interventions we can perform on the dynamics in an experiment, how we make those interventions, and even how we identify processes within an experiment. These points are discussed in more detail in a later chapter.

It is worth noting here that much of the robustness literature makes no claims about the ontological features of robust entities within scientific experiments.

---

62 See Norton (2003, 2010, 2021). Norton's material theory of induction rests on background facts to provide for the strength of an inductive inference. Similarly to Norton's account of replicability, I believe the material theory of induction is co-optable by the process realist because these background facts will be about processes. However, Norton's account is nowhere explicitly process-realist, nor does it need to commit to any realist position.

63 It is important to note that this absolute stability is what the thing realist needs to justify the existence of things. A static thing cannot be the sort of entity that is relatively unchanging. Such an entity is still dynamic, and so is not static. A thing must be unchanging full stop.

Their arguments are focused more on the epistemology of robustness reasoning. Insofar as such reasoning is still an important part of science, I endorse it. Robustness arguments are co-optable by the process realist, after all. For this reason, I have refrained in this section from criticizing specific authors in the robustness literature; their work is not being called into question except where robustness is used for making explicit ontological claims.[64]

However, equally of note, even in a literature focused almost entirely on the epistemology of scientific practice, the participating authors make implicit ontological claims. This is seen not through their explicit arguments, but rather through the examples of robustness to which they refer. Chief among these are the examples of Brownian motion and the confirmation of the molecule, an example referred to almost universally in the literature. The example itself is taken as an historical moment in which atomism was confirmed and plenum theories refuted. W. Salmon (1984) makes the thing-realist tint of this explicit.[65] Even when robust phenomena are not explicitly thing-realist, as in the case of the conjunction fallacy studied by Crupi et al. (2007)[66] or the Volterra principle discussed by Weisberg and Reisman (2008), the discussion of these phenomena seems to inherit the orthodox assumptions of thing realism anyway. Firstly, those robust phenomena that do not conform to a thing ontology (as does the robustness of the molecule) are treated as a type of "model robustness," and not a type of robust entity within the world. I.e., the Volterra principle and conjunction fallacy being as they are not explicit substantial things, they are robust epistemic tools for modeling what is real, not real themselves. Secondly, when such modeling tools are reified, they are reified in terms of thing realism, as in the case of climate models.[67] This is seen even more explicitly in the philosophy of science literature surrounding robustness. One key factor in robustness explanations, according to some, is the "ontological independence" of the lines of evidence. This ontic independence is exclusively thing-ontic, since independence is prohibited for any entity that must be contextually identified (such as processes).

I will spend a fair amount of time refuting the supposed triumph of thing realism—the Brownian motion case—in Chapters 3 and 4. However, this refutation and all other refutations of robustness I have offered come from the lamentable fact that thing realism is so firmly ingrained in orthodoxy that few stop to consider

---

[64] Eronen (2015) is perhaps the best example of this.
[65] See also Cartwright (1983, 1991), Mayo (1986), and Hacking (1984) for further discussion in this line.
[66] See also Crupi et. al. (2008).
[67] C.f., Lloyd (2010) and Parker (2011), who both use the thing-realist language of states and state-causes.

that a slight ontological shift could offer a solution to the problems of robustness (i.e., underdetermination, typological problems, etc.). In addition, while some in the robustness literature will no doubt appeal to the non-ontological character of their investigations, even if this were true, the epistemic/descriptive versions of robustness are equally problematic, as I will discuss in §2.2.2.

2.2.1.4 Summary of Experimental Stability Underlier Arguments

Each of the underlier arguments presented in section 2.2.1 were based on stabilities *observed* in experiments. As such, they formed the strongest empirical arguments for the existence of things to underlie processes we could posit. However, in each case, these underlier arguments relied upon a dubious premise: that stabilities observed in the world entailed the existence of statics in the world. Deductive attempts to justify this failed because no non-question-begging principle strong enough for the deduction could be produced. Inductive attempts fair better, but ultimately fail because they would require assumptions about contingent features of the world that are unverifiable. Both deductive and inductive versions fail in part because of known physics: in order for us to deduce or induce that there is a lack of change, we would need to show that there is a threshold past which it is impossible to supply additional energy to probe and perturb a system. Further, there would have to be a stable, persistent entity that persists through perturbations at that energy threshold. This, at least, has not been (and is unlikely to be) observed. Lacking this, there is always an amount of energy that can destabilize any stable entity posited in any theory or as part of any experiment.

Crucially, if stability does not entail staticity, this *opens the door for viable and preferable process-ontological interpretations of stable systems*. The thing realist doesn't technically need entailment. All she does need is that a static account of stability is preferable for explanatory goals of science and metaphysics. However, since there is no strong entailment between stability and staticity, the process realist is at liberty to reinterpret stability claims in terms of processes. I have already offered suggestions of how to do this, leaving the thorough discussion for §2.3. However, the general strategy is the following: all we need to act as stable underliers for experimental dynamics are processes that are characteristically more stable than the experimental dynamics, or are unperturbed by the experimental interventions. Stability claims are made true by referencing this loose hierarchy of relative energy and time scales in addition to the specific physical dynamics involved in the relevant systems.

2.2.2 Stability of Descriptions and Models

Our second broad category of underlier arguments is built on claims about the stabilities present in one or more of our scientific models: stabilities of descriptions. This stability of description follows a similar pattern as the stability of experimental systems we have already discussed. I.e., the stability of description comes in three forms.[68] First, in every description of an experimental system, certain terms appear that supposedly refer to static entities, i.e., nouns. This is similar to the appearance of persistent aspects in every experiment. Second, our descriptions, and especially the general terms that appear in those descriptions, are consistent in their application. This is similar to the stability we observe in repetitions or reproductions of a single experiment. Third, across multiple different descriptions of various systems, it is not uncommon for certain similar terms to appear, apparently referring to the same entities in different systems. This is similar to the robustness we observe between different experimental systems.

Just like with underlier arguments based on experimental stabilities, those based on descriptive similarities come in three types, following (roughly) the three types of descriptive stability. However, these descriptive underlier arguments are generally weaker than those based on experimental stabilities. In large part this is because arguments based on stabilities of description are implicitly reliant on the stabilities found in experiments. When they are not reliant in this way, descriptive stability underlier arguments rely on assumptions about the connection between language and metaphysics. Those arguments that implicitly rest on the arguments we have already discussed will prove easy for the process realist to refute or co-opt. The latter type of descriptive stability arguments—those that rest on quirks of our language—are also quite easy to refute or co-opt. This is because process realism does not require that we alter our language, only that we recognize that language (especially mathematical language) is not indicative of a particular metaphysical entity. In other words, the arguments discussed in this section fail to rule out that the underliers are processes and not things.

2.2.2.1 On Nouns
2.2.2.1.1 The Characteristic Underlier Argument
In many models or descriptions of experiment(s), there appear terms that apparently demarcate static things. These terms are, typically, nouns: terms like "electron," "organism," "chemical," "molecule," "galaxy," and so on. These terms are

---

[68] One might expect this, since descriptions of experimental systems should align with the goings on in those systems.

not inherently referring to changes in the world—they are not verbs or adverbs—and so they must at least be stable entities. The intuition, then, is that these terms are not merely referring to something stable, they are referring to something static, i.e., a thing.

This trades on the standard, so-called inferential approach within analytic ontology.[69] The program of the inferential approach is to interpret ontology as a properly simplified domain of linguistic reference and inference. In other words, whatever our sentences (minimally) refer to as truth-makers for the range of meanings we can generate are necessary ontological posits. Our ontology, then, posits entities that we check against our linguistic data. If "we would say P of X," then our ontological posit, meant to act as an interpretation of that sentence, should be such that we are correct in predicating P of X. If it fails to allow this, the ontology fails to capture some linguistic data, and is thereby made less preferable than some ontology that does capture this data. The strength or success of an ontology then is the sum of the successes it has in capturing our functional uses of referring terms in everyday (or technical scientific) speech.

In the context of scientific language, the intuitions behind this argument are strengthened by assumptions made about mathematical reference. In a mathematical model or description, there is a much clearer demarcation made between terms or functions that vary and terms that do not vary. E.g., the mass factor in a kinematic equation is an unchanging factor, as is the number of molecules per mole in a thermodynamic system (i.e., Avogadro's number "N"). In contrast, terms such as the position of that mass, or the volume of a mole of gas undergoing adiabatic expansion are both described mathematically as functions of time. This suggests, more strongly than the distinction between nouns and verbs, that there are entities in the world that do not change, simply because they are referred to by terms that lack a dependence on a temporal variable.

In other words, we have:

(Noun Underlier Argument):
($P1_N$): There are noun-terms in our theories and models (more generally, descriptions) of experimental systems.
($P2_N$): Noun-terms admit of the sort of inter-theoretical/inter-descriptive inferences typified only (or most often) by entities with definite, non-contextual properties, most especially quantities (stability entails staticity).

---

[69] The inferential approach is built on the ideas of Carnap (1934) and Quine (1960), and has become multifarious in its specifics over the last century. For reconstructions of the history of this approach, see Seibt (1996, 1997, 2000).

(P3$_N$): We should adopt as part of our ontology/realism about scientific models only those entities that are necessary to make true the inter-descriptive inferences we make in and among those models (inferential approach to ontological modeling).

(C$_N$): Therefore, there are entities with definite, non-contextual properties/quantities, which is just to say that there are things.

The mathematical intuitions behind this argument are apparent in the best example of it in philosophy of science. Namely, in accounts of scientific reference. There are two predominant approaches: the semantic account (Da Costa and French 1990, 2003; French 2016; van Fraassen 1989, 2014;[70] Suppe 1989) and the syntactic account (LeBihan 2012; Frigg 2006, 2010; Goodman 1977;[71] Halvorson 2012).[72] In both a metaphorical isomorphism is used to suggest that all terms within a theory or model that bear a one-to-one (and onto) relationship with the mathematically definite features we observe in experiments must be real or referential. More simply, if a measured value can be produced in an experiment, and a model mathematical term can take this value, we have evidence of real reference. The difference between the semantic and syntactic accounts then arises as a result of whether this syntax is taken as enough to define the reference class of a theory (the syntactic view) or whether the reference class is further determined by inferential use within the context of the model (the semantic view). French describes the key difference like this:

> Given, then, that scientific models are, primarily, representations, in what sense may they also be mathematical structures in the way that the semantic approach proposes? The answer is straightforward: 'A model is a mathematical structure in the same sense that the *Mona Lisa* is a painted piece of wood.' In other words,, both the representational content of the painting and the action painted piece of wood are what make the *Mona Lisa* the artefact that it is, and similarly, there is more to a model, as a scientific artefact, than the relational structure [syntax] in terms of which we can define embeddability, isomorphism and so on. (French 2016, 4)

I.e., scientific theory representation is the result of both a direct isomorphism between a theoretical term and an entity in the world *and* a context of use on the semantic account. Thus, we infer that a theoretical term refers in virtue of both

---

[70] See Suárez (2010) for a description of how van Fraasseen evolved from a proponent of the syntactic approach into a proponent of the semantic account.
[71] Described extensively in Polanski (2009).
[72] See French (2016), Frigg (2006), LeBihan (2012), Suppes (1989), and Halvorson (2012). See also Glymour (2013) and van Fraassen (1989, 2014).

its direct mathematical relationship to experimental outcomes and the inferences we draw within the model about those theoretical terms. This is almost identical to the ontological approach in the inferential program in ontological methodology.

Importantly motivating the semantic theory of scientific reference is the issue of equivalence between theories. The problem is that two theories with different terms (e.g., Heisenberg matrix mechanics and Schroedinger wave mechanics, Legrangian and Hamiltonian formulations of classical physics) can sometimes be shown to have equivalent empirical support and import. In such cases, the semantic theory makes use of the less-metaphorical isomorphisms one can produce between the two theories to show that they are indeed equivalent. The intuition, then, is that there must be a reference relation from theory to world that is preserved under this isomorphism. The natural candidate is another isomorphism. See, for instance, Halvorson (2012), who argues for exactly this in order to suggest that the syntactic identity between such equivalent theories suggests that they are the same theory (in terms of referential domain). See also French (2016) for a response to Halvorson on this point.

In both the semantic and syntactic views, we find arguments that theories refer to things. The syntactic account takes the mathematical features of, e.g., certain scalars in scientific models as indications of entities with definite, scalar quantity values in the world (i.e., things). The semantic account, acting as a hybrid of the descriptivist and intentionalist models of general reference, that the descriptive and intentional elements of a model or theoretical term force its relationship to specific types of entities in the world. I.e., thing-like terms must refer to things: "molecule" must refer to an object with static "molecular properties," because "molecule" is *used* to refer in this way. Such terms are nouns with descriptive adjectives but no apparent natural adverbs. Thus our inferences using theoretical terms force us to adopt thing-ontological claims. Examples of both thing-inference can be found in Gooday and Michell (2013), Fraser (2011), and North (2009) for classical, field-theoretic, and general physical things respectively.

2.2.2.1.2 The Refutation

There is a problem with the standard account: while the intensional inferences of a linguistic structure may provide many easy cases in which grammatical forms are apparently indicative of ontological types, there are also many cases in which this does not hold. For example, one possible inference I may draw about ontological types from grammatical forms is that nouns (and therefore noun-referents) can be predicated of. I might therefore infer that noun-referents are the sort of entity that can admit of definite properties like "being square," corresponding to the predication of squareness of various nouns in sentences like "the table is square."

However, there are also nouns that admit of both definite and indefinite predications. There are nouns like "run," for which both the predication of "long" and the predication of "3 miles long" are grammatically concordant. Moreover, there are nouns (gerunds) that admit also of adverbial predication, such as "running." Are we then to suppose, strictly following the inferential approach, that noun-referents are general enough to admit of definite, indefinite, and adverbial properties? Such an entity is not clearly a thing anymore, nor is it clearly a process.

More importantly, many of the examples above refer intuitively to specific ontic types because of the way in which we employ the terms. "Table" is intuitively referring to a thing. "Run" is referring to a complete process. ʳRunning" is referring to an ongoing process. Three nouns, with three different ontic types. The existence of nouns alone cannot be indicative of things. We must look further to the function and use of the term in actual reference instances. Specifically, how do we use noun-terms in experimental descriptions?

The answer is that we do not consistently use nouns in our experiments to refer to any particular ontological category. Indeed, the reference of a term is so highly contextualized to the experimental scenario that exactly the same term can admit of *contradictory* predications. I go through an extended example of this in Chapter 5 on nuclear modeling. Suffice it to say, for now, that terms like "nucleus" can support predicate sentences like "the nucleus has internal energy structure" and "the nucleus has no internal energy structure" depending on the experimental context of the relevant models.

We should also be dubious of this argument from the outset because of its implicit cultural and historical contextual assumptions. Seibt (2015) has argued that the inferential approach needs to be modified to be sensitive to the differences between Indo-European and non-Indo-European languages. Moreover, she argues with reference to recent work on the "Linguistic Relativism Hypothesis" that Indo-European languages predispose speakers towards an ontological prejudice for thing-like noun-referents. This, coupled with her modified inferential approach, leads her to argue that general, subjectless processes can act as the sole ontic type to encompass the linguistic needs of a non-Eurocentric linguistic analysis. I.e., processes alone can satisfy the needs of reference. This proposal combines the scarce attention to subjectless processes in ontology with insights from the linguistic field of aspectology, as adumbrated by Mourelatos, but does so by including typological aspectology. [73]

---

[73] See also Vendler (1957), Kenny (1963), and Zemach (1970), who respectively provide a distinction between classes of referents for English verbs in terms of action types (Vendler and Kenny) and introduced a notion of "process" that admitted of ontological categorization (although without the subjectlessness).

A simple example of this will suffice to drive home the point. English being verbs (or rather the lexical semantics of the nouns paired with these verbs) seem to suggest in their use that whatever is is definite and thing-like. "I am a person," "She is moral," "Trees are all plants," "We are a community," "Molecules are structured." In each, the definite verb seems to define the definite existence of a thing-like entity. The nouns all apparently define the predicates we would attribute to the objects in the world characterizing their differences as different definite entities—trees, I's, She's, We's, and molecules—(Quine, 1953, 13f.). Because these nouns all provide a definite categorization of the thing that exists, is a person, is moral, is a plant, etc., we typically assume that nouns demarcate definite objects. However, the being verb in Chinese Hanzi and Japanese Kanji indicates no such definitiveness. Indeed, the Kanji 是 is a pictograph composed of the radicals for "sun," "below," and "movement." The Kanji, then, is used to indicate that being is "the change that underlies everything under the sun." Existing things are defined by change, and by participating in the whole of the world. It is this recognition that allows us to interpret the famous (simplified) Sutra: 見山是山, 見山不是山, 見山秖是山 (roughly: "mountains are mountains, mountains are not (不) mountains, but mountains are merely (秖) mountains). [74]

The fact that a processist account of the referents of nouns across languages has already been given (Seibt 2010, 2015) should prove as sufficient response to the nouns argument. Once more, it is enough to show that processes can meet the inferential implications of nouns equally well as things. This entails that the nouns argument cannot rule out that the underliers of our noun-terms are processes rather than things. Moreover, once again, there is some reason (cultural, historical, and scientific) to suppose that processes will be better ontological commitments in the domain of interpreting language usage. This is both because processes can meet the contextuality needs of inter-cultural analysis of language, and because processes can allow for consistency of reference where things cannot (cf. Chapter 5 of this work).

Perhaps more interestingly, noun arguments for things can be co-opted as arguments for processes using the same methods from ontology and philosophy of science. The semantic account of model-reference is particularly useful to the process realist, since it naturally demands that scientific models are partially defined

---

[74] The Japanese Kanji+hiragana version of this looks similar in that the Kanji are pictorially the same as the Hanzi in the Chinese version of the sutra. The meanings of the Kanji and Hanzi are similar, although the contexts of use for these characters differs enough to leave interpretation somewhat open. The section I present above is simple enough that this philological issue poses little issue.

by the context of the inferential use in the practice of science.[75] While I would not commit wholeheartedly to, e.g., French's (2003b, 2016) partial structures approach, and would instead follow the spirit of Seibt's (2015) program by requiring alteration of the methods of the semantic approach, it is interesting to note that French's partial structures account was co-developed by Bueno (Bueno 2000, 2016; Bueno, French, and Ladyman 2002; Bueno and French 2011, 2012, forthcoming). Bueno has long been process-realist-adjacent at least (2000, 2019), which might suggest that a purely process-realist account of scientific modeling practice and reference will eventually see success, much like the success of process ontology in accounting for linguistic practices.

2.2.2.2 Consistent Reference and "Fine Tuning"
2.2.2.2.1 The Characteristic Underlier Argument

Our next underlier argument comes from the stability of our descriptions in their application. I.e., our (best) descriptions are consistent in their application between different particular systems of a similar character. For example, when I point at a tree and describe its size or shape, "tree" refers in much the same way as when you point at a different tree to describe its different size or shape. In other words, regardless of the particulars of the tree I am currently referring to, the word refers to the same sort of entity as the one you are referring to when you use the same word. This suggests that there is a stable set of properties of the entity in the world (whatever it is) that allow us to consistently refer to it. The argument then goes that a stable set of properties can only exist if that stable set is static, and static properties can only be attributed to and exist within things.

We should note at the outset that this type of argument follows a similar pattern as the arguments for things offered by the semantic account of scientific reference. Namely, the patterns of our inferences entail that there is some stable collection of referent things. Nevertheless, the argument is slightly different from the noun argument above, as I show.

This argument is immediately plausible in the mathematical context of scientific modeling practice. A gas, for example, is described thermodynamically using a set of core properties: typically, osmotic pressure and molecular (or atomic) weight. These properties are given quantitative definitions, units of measurement, and conventional scales for comparison with different gases. Then, we identify a particular gas—e.g., Helium gas—by its particular core properties. This lets us apply our general models for gas expansion and contraction to Helium gas consistently and precisely.

---

75 See Suárez (2003) for an argument that representation in science is contextual in this way.

Crucially, the existence of stable, determinate, identifying features of the gas is what allows us to reidentify it in each experimental or theoretical context. It is because Helium has a determinate atomic number (=2) and molecular structure (monatomic) in every thermodynamic context that it can be identified wherever it appears in models, theories, or experiments (at least, so goes the argument). Thus, we have:

(Consistent Reference):
($P1_{CR}$): We refer consistently with a term in multiple (similar) contexts/models.
($P2_{CR}$): The term's successful reference is picked out by identifying stable determinate features of a system.
($C1_{CR}$): Therefore, by induction over multiple similar identifications of successful reference, there is a collection of stable, determinate features of systems that is context invariant.
(C_{CR}): Therefore, there is a thing (defined by this collection of stable, determinate features).

This argument essentially seeks to show that the existence of determinate features to which we refer consistently entails a subject for those features. The induction contained in the argument is an induction over cases of identifying these features, e.g., instances of determining the atomic number of Helium, or the internal energy structure of an atom's/molecule's electrons. The more instances in which we successfully determine these features, the more likely it becomes that those model features have real referents. These real referents must then be as stable and determinate as the values/variables we see in our models.

This is exemplified in the literature on essential properties. Three lines dominate this discussion: the modal account of essential properties (standard and nonstandard), the definitional account, and the intrinsic/extrinsic account. The details of these are not important for our discussion, but the literature in which these accounts exist is rich with examples of the consistent reference argument as I have called it, as well as implicit underlier arguments of other sorts. Fine's (1994) definitional account is a prime example. Essential properties, according to Fine, are those that define the state of being of a thing. They are therefore properties that consistently define the thing in every context, and to which we implicitly refer when we mention the thing. Zalta's (2006) "encoding" account is similarly suggestive. According to Zalta, an essential property is one that encodes the object in which it is found. I.e., reference to the property is implicit reference to the thing in which that property is found. In both of these, it is the consistency of reference to specific properties that defines a thing. An important contribution to this literature comes in N. Salmon's (1979, 1981, 2003, 2005) work which makes the link between consistent reference and the essences of things more explicit. In addition, it should be noted that the history of Aristotelianism is ever present in this literature, especially the connection offered by Aristotle between the definition of an entity and its ontological essence (see Peramatzis (2011)).

2.2.2.2.2 The Refutation

While the consistent reference argument is suggestive, the process realist has the now-standard response. Namely, this argument does not rule out that the consistent referents of terms and features of our models are processes. Given that processes can be given identifiable features (Seibt 2010, details provided in §1.2), there is no reason to suppose that processes cannot be identified and reidentified consistently in scientific experiments and models such that we can consistently refer to them. Instances of such reference abound. For example, we refer to motions both in natural language and mathematical language consistently. We also refer to interactions, e.g., "gravitation" or "electromagnetic repulsion" both in natural language and mathematical language. These entities are dynamic entities, identified by dynamic features such as dynamic shape, dynamic context, measurability, determinable-ness, and so on.

What is more, historical examples exist in which a model of a system refers to a real interaction or motion that was as-yet unobserved. For example, the model of the nucleus first proposed by Heisenberg in 1932, and later made empirically adequate by Yukawa, proposed that the nucleus was held together and made stable by an exchange force interaction between protons and neutrons. This interaction was quantitatively incorrect in Heisenberg's model, but was later found and explained by Yukawa.

The thing realist may claim in response that neither motions nor interactions can be referred to without things. E.g., motion must be the motion of something, and interactions must occur between two (or more) things. This is almost exactly the point made in Strawson's (1966, especially part 2, chs. 1, 2, and 3) argument that reidentifiability is impossible in a purely processual (or event-based) metaphysics.[76] According to Strawson, processes (and events) need things to give them reidentifiability in part because only things can enable coordinate descriptions of space and time. Thus, when we go to reidentify a system from earlier or elsewhere, we need to be able to locate that system in space and time, and so require things.[77] However, this would be a clear instance of overreach. There is no reason to sup-

---

[76] Strawson's argument in part 2, ch. 1 is that reidentifiability is the result of existence in a spatiotemporal nexus.
[77] Seibt (1990, 27–37) argues extensively against Strawson in particular.

pose that the underliers of motions and interactions must be static if processes have identifying features independent of having a subject.

The consistent reference argument can and should be co-opted by the process realist. For example, interactions are indeed interactions between two or more entities, and it is true that we refer to an interaction consistently in virtue of similarities between the interacting entities in addition to similarities in our description of the interaction itself. More generally, it is true that similarities between referenced entities are demarcated by stable features of those entities.

However, stability does not entail staticity; there is no reason to suppose that similar entities are things simply because they are stable and similar. Echoing Fine (1994)'s definitional account of core properties of things, we might say instead that there are core processes to which we refer whenever we use certain terms. These core processes would need to be *stable enough* to allow for us to identify the less stable interactions, motions, and (generally) dynamics that we observe in an experimental system. For example, the terms "the nucleus" or "the molecular bond" are not used to consistently refer to static things, or to a core set of static properties of unknowable substances. Rather, they refer consistently to a collection of dynamics with characteristic energies and relative stabilities, and to particular classes of interactions. These examples are discussed in greater detail in later chapters.

I contend in later chapters that we not only can, but should and do refer to processes as consistent identifiers of experiments and of models. Indeed, I argue that the substitution of dynamic referents for thing-posits is at the core of the historical development of physics. Namely, we begin by positing a stable thing because of dynamics observed in a system. We then show how this thing, and all of its stable properties, are no more than dynamics themselves, albeit dynamics that are more stable than those observed in the original experiment(s). If we are so inclined, we then posit further things to underlie these new dynamics, only to continue the process and discover how these further things are collections of dynamics themselves. In short, while thing terms may be useful, they are only ever placeholder terms for further processes. At best, a thing is never referred to. The term is only a linguistic tool for arbitrarily demarcating the processes of interest in an experiment. I will show this in the context of a simple thermodynamic model, and again in the context of a more complex thermodynamic argument famous in the literature for supposedly offering support for real things. That is, I will consider Perrin's (1909) argument for atoms/molecules in thermodynamics.

2.2.2.3 Unification of Models
2.2.2.3.1 The Characteristic Underlier Argument
The last argument we will consider that makes use of stability of reference and description rests on moves throughout history to unify disparate models and descriptions of experiments. Suppose we have multiple descriptions of different experiments, and that in each of these descriptions, the same term appears. This suggests that, whatever this term refers to, that referent entity is the same in each of the descriptions. Since each description will (inevitably) involve some reference to dynamics and change, this referent entity is unchanged in each of the described situations, the experimental dynamics. Thus, ultimately, the referent entity is unchanging, because only then could it act as a unifier of our many and varied descriptions of many and varied experiments.

This sort of unification should be differentiated from what we saw in the robustness argument above. Namely, the difference lies in what is meant to act as the unifier. In the robustness argument, we unify our experiments by presuming a common ontic element within those experimental systems. This argument, however, attempts to unify descriptions of experimental systems by presuming a common referential structure for those descriptions. Given that any ontological conclusion drawn from the existence of a common referential structure will likely involve positing a common referent, and so a common real entity, unification arguments and robustness arguments are oftentimes inextricably linked. In particular, many of what I call unification arguments will ultimately rely on what I call robustness arguments in order to justify their central claims.

However, just like the robustness argument discussed above, the unification argument attempts to justify the existence of a static thing with an induction over many descriptions. The weakest version of this induction is an induction over similar descriptions of similar experimental systems. For example, let us say we describe one particular nuclear experiment in terms of a nucleus qua collection of colliding nucleons, and another in terms of a nucleus qua collection of quantized nucleon energy shells. The (weak) unification argument would argue that, despite differences in the theoretical description of these two "nuclei," the fact that there are unifying features (e. g., the existence of nucleons, the exact number of protons and neutrons in a given nucleus) as reason to suppose that the term "nucleus" can unify the two descriptions.

More powerfully, the thing realist could attempt to show that a unifying concept of the nucleus is in operation in both descriptions. Finding such a unifying model, that successfully encompasses the explanations and descriptions offered by both models within the same theoretical structure, would act as justification for supposing that this unifying concept is real. Describing historical instances of unification arguments such as this appear as a core research program in Janssen's work (delivered most evocatively in a talk entitled "Arch and Scaffold" in 2015, see Janssen 2004a, b, 2011).[78]

What we get is an argument that whatever core term or structure responsible for unifying two previously disparate domains of reference/representation must be real. Most often, the unifying features of a theoretical domain are taken to be the thing-ontic descriptions of the relevant system, as in Linnaeus's unification of biological classifications resulting from core biological properties, or the unification of Heisenberg and Schrödinger mechanics, or the unification of various thermal dynamics supposedly present in Perrin's (1909) argument for atoms (to be discussed in Chapter 4). This gives the following argument for things:

(Unification Argument):
($P1_U$): Every model and theory has a domain of reference.
($P2_U$): In (certain cases of) two or more disparate experiments, we see the emergence of stable or robust referential portions of our models of those experiments.
($P3_U$): The stable portions of the domain of reference are referred to by the terms identifying things, i.e., the terms that identify the essential and stable determinate features of the experimental system(s).
(C_U): Therefore, the referential unifier of these disparate experiments is a thing or thing-like entity (it has determinate properties, it manifests determinate structures).

($P3_{U1}$) is doing all of the work of this argument. Implicitly, this premise is reliant on an induction over many cases of referential similarity in many experiments. E. g., the stable features of the nucleus—those that are not involved in or responsive to the perturbations we perform in our many experiments—are the only possible stable aspects of our descriptions of the nucleus in each experiment.

This should be rather intuitive, even to a process realist. The fact that each different experiment (and therefore each description of it) involves a different set of dynamics seems to entail that any two experiments will differ because they involve different dynamics. The fact, then, that there are similarities between the descriptions—e. g., we describe both nuclear experiments as experiments on *the nucleus*—is highly suggestive that this similarity cannot be because of the processual descriptions of the system. Those processual descriptions, e. g., descriptions of the interventions performed on each nuclear system, are not unifying by default.

---

[78] See also Kitcher (1981) for one of the key accounts of explanatory unification.

A special class of unification arguments are the ones seeking to unify all of science or all of physics under a single umbrella ontology. These arguments tend to be explicit in their preference for thing ontologies. For example, an explicit preference is placed on the reductionist ontology of fundamental particles as the unifying ground of all scientific phenomena in works such as Bennett (2017) and Sider (2011), both of whom combine the notion of a complete minimal basis[79] of fundamental entities with this preference for a particle ontology to form their complete ontological picture.[80]

These special "universal" unification arguments are not terribly relevant to our discussion. They rarely rest their unification premises on actual descriptions provided by experimental models, opting instead to import various reductionist principles applied to theories, rather than experiments, to support their move to universality. Following the program of Chapter 1, the goal of this chapter is to show that the thing realist cannot infer *from specific experiments and models of them* to the existence of things, unlike the process realist. Whether or not things provide a universal unifier is moot.

2.2.2.3.2 The Refutation

By now, the strategy of refutation should come as no surprise. The process realist needs only to show (a) that the unification argument cannot rule out that the relevant underlier (the unifying descriptor) is itself a process, and (b) that we can co-opt the unification argument to argue that the proper unifiers of different model-descriptions of experiments are indeed processes. This is as simple as exhibiting such an instance, and noting that it is likely to generalize.

Just like with the robustness argument, the inductive risk shouldered by the proponent of the unification argument is too great. There is no term that refers consistently under all descriptive and experimental differences. As such, the best the thing realist can hope for is to provide a term that refers consistently in almost all contexts. Such a term can easily refer to a process; the stability of the term does not entail the staticity of the term, and therefore the term's referent

---

[79] See Bennett (2017), Jenkins (2013), Paul (2012), Raven (2016), Schaffer (2010), Sider (2011), Tahko (2013, 2014), and Wilson (2012, 2014, 2016).

[80] See also Morganti (2009), Ney (2015), Wolff (2012), and Zimmerman (1996). It is interesting that even in diverging from standard particle-ontology views, authors like Wolff (2012) (who advocates structuralism) and Zimmerman (1996) (who advocates for "pure substance" or gunk) still commit to a thing ontology in order to characterize their fundamental layer of reality. Something similar can be said for Schaffer's (2003, 2004, 2009, 2010a, 2010b) holistic monism (see also Ismael and Schaffer (2020) for what I take to be an interesting new movement on Schaffer's part toward a more process-oriented ontology).

need not be static either. Thus, the thing realist cannot rule out that the unifiers of our descriptions (the stable terms present in our descriptions) refer to processes and not things.

I note here that the unification argument, properly applied to historical examples, is actually an argument for process unifiers. That is to say, the historical unification of disparate, incompatible descriptions of experimental systems has been accomplished primarily by eliminating the unnecessary thing terms from the unified model while retaining the relevant processual aspects of the unified description. This is most obvious in the unification of models of the molecular bond (see Gavroglu and Simões 2012, especially ch. 1). I provide an extended and novel discussion of a similar unification of nuclear models in Chapter 5, of particular interest because these models have long eluded unification in thing interpretations.

Moreover, we can make clear why unification should always come by rejecting things and retaining processes. Namely, unifying models must always retain the descriptions of experimental systems. Since, in Chapter 1, we have seen that such descriptions essentially include descriptions of processes (interventions, dynamic responses to interventions, interactions with observers, etc.), this means that unifying models must at least describe all of the relevant experimental dynamics of the unified models. Thing terms, in contrast, are unnecessary.

A little investigation reveals that unification arguments can offer very few obvious examples for the thing realist. Most instances of unification present in the literature are actually instances of unification of method, of mathematics, or explicitly of processes and dynamics. While the rare instance of apparent "thing unification" exists (cf. the atom and the electron), some of the most prominent instances of historical unification are unifications of processes. Newtonian gravitation, which unifies earthbound and celestial gravitational processes, is one such example. The categorization of oxidation and reduction reactions under a single archetype in the chemical revolution is another. To think, therefore, that unification is something only available to the thing realist is mistaken.

2.2.2.4 Summary of Descriptive Stability Underlier Arguments

We end our analysis of descriptive stability underlier arguments noticing that they fare similarly to those based on experimental/observed stabilities. All of them begin with a sound premise: the positing of a particular sort of stability. All of them then make the case that this stability can only exist if something static exists, by either inductive or deductive means. But ultimately, all of them commit the same mistake as the experimental stability underlier arguments: assuming that stability entails staticity. In fact, the existence of stability only entails that there are more and less stable entities.

However, descriptive stability underlier arguments differ in that they rest on more complicated philosophical work on the relationships between utterances and their truth-makers. Our analysis of these arguments, is therefore much less easily confirmed and supported than were the analyses of the arguments in §2.2.1. In each instance, I argued that such referential and inferential networks can indeed be coordinated around processes and not things. However, as noted above, the fully developed account of scientific representation using process ontology is still some ways from complete.

That said, process ontology already has a significant history of work in reference and language, as I have pointed out. We can therefore be somewhat sure that process ontology can indeed provide a successful account of the reference of scientific models and terms. I will spend some time in Chapters 3, 4, and 5 showing examples of how to make use of this existing process-ontological architecture to interpret various historical and contemporary scientific models.

2.2.3 Manifest and Assumed Stability

In §2.2.1 and §2.2.2, I argued that underlier arguments based on experimental and descriptive stabilities failed as arguments for static entities, i.e., things. Given that we have now largely ruled out that things can be (safely) inferred from either observed or described stabilities, one might wonder whether there are any underlier arguments left to be offered. In fact, there is one final class of underlier arguments left to discuss: arguments from assumed manifest stability. Namely, one might think, generally, that since the world is a material world, it is therefore composed of a substance—matter—that is thing-like (defined by atemporal, determinate properties).

While the basic argument is simple, this category of underlier arguments is the most varied of the categories we have discussed. These underlier arguments trade on deeply embedded assumptions about metaphysical priority relations. For example, the assumption that everything in the world must be material plus the assumption that matter is a substance entails that everything in the world must ultimately be composed of substances (things). Other examples include assumptions about what can act as a common cause or unifier, the proper definition of change, what sort of entity can bear symmetry, etc. As such, these arguments tend to be quite convoluted.

That said, I will be discussing these arguments mostly for the sake of completeness; they are largely beyond the scope of this chapter. The argument of this chapter has been to show that facts about specific experiments, observations, and experimental descriptions cannot support, on their own, inferences to things. As such, any argument for things that takes as data facts and features of experience or metaphysics or language that go beyond what can be found in manifest experiments will not be immediately relevant here. Suffice it to say that these arguments will largely face the same challenges as the arguments we have already discussed. Namely, they will fail to rule out (without quite strong and dubious assumptions) that the relevant underliers are not processes.

2.2.3.1 The Argument that Matter Is Essentially Substantial

We begin with a rather interesting argument. Namely, that matter physics is fundamentally concerned with discovering the fundamental *definite* components of the world that ground all relations, properties, and dynamics of higher level systems. I.e., that matter physics is about discovering that which is "not predicated of a subject but [of which] everything else is predicated" (Aristotle, *Metaphysics* 1017b1). This is readily apparent in the debates about field vs. particle interpretations of quantum field theory, and the philosophical discussions drawing on or referencing this debate.

I say that this argument is interesting in part because it contains within it a host of historical and philosophical assumptions. The philosophical assumptions of the substance paradigm that underlie this argument have been cataloged already (see Seibt 1990), in which Seibt catalogs 22 assumptions or components of an ontological commitment to "substance." This allows her to argue that claims involving substance face various challenges because of the specifics of the sub-collection of assumptions involved. Given that she has made it part of her research program to dissect and defeat various thing-ontological claims in this way, I will simply reference this work, and focus primarily with the historical point: that matter physics (has been/is currently) interested in producing thing-like referents to underlie and define the concept of matter and its properties and dynamics. I will present only a brief overview of the history here in order to set the stage for the discussion to come in Chapters 3 and 4. A full account of the historical development of matter physics must be left for another work entirely.

The history of matter physics is at least as long as the history of philosophy. It begins with mythological accounts of creation and the nature of the world. In these, we find consistent reference to some sort of "stuff" (usually primordial water) out of which some agent or force created all entities living and dead. This informed the pre-Socratic Milesian monists, who began to ask the explicit question of what (stuff) was the fundamental component of all (material) entities in the world, motivated by their observations and accounts of what could change into what. For instance, Thales postulated that water is the fundamental stuff, noticing that in many material transitions water is a common factor. From these debates, Parmenides and Heraclitus developed new accounts of how to organize the material systems of the world and how to understand change itself. Together, they formed the first dichotomy between what might broadly be considered change-centric (Heraclitus) and changeless (Parmenides) accounts of being.

It was at this time that matter and motion became associated in physics. In particular, the ancients posited first and foremost in their physics a strong explanatory connection between the matter of a body and its natural and possible motions. This is seen in both Plato (*Timaeus*), in which we also see a clear expression of the Heraclitus-Parmenides distinction ("What is that which always is and has no becoming; and what is that which is always becoming and never is?" *Timaeus* 27d5),[81] and in Aristotle (*Physics II*). The key was this: a body could move in a particular way (naturally) if and only if the body was composed of the appropriate matter. Thus matters and motions (especially circular vs. linear motions) were co-explanatory. This same explanatory connection is apparent in the medicine of the time as well, which posited a strong explanatory connection between the matters of the body and its functions or natural motions (see "Waters and Airs" in the Hippocratic Corpus).

Much of the history leading to the scientific revolution is a story of increasing precision and calculational acumen. However, this fundamental connection between types of matter/substantial stuff and physical motions remained largely unchallenged until the time of Galileo (it was also apparently challenged in Hypatia of Alexandria's work, but no primary sources remain to corroborate this). Galilean physics divorced the quantity of motion from the quantity of matter, and to some extent the specific substantial constitution of matter, by noting the identical free-fall motions of different bodies.

However, while an important development in matter physics, this did little to unseat the prevailing interpretation of matter as a kind of stuff similar or identical to the philosophical concept of substances. Cartesian physics shows that this assumption was still largely unchallenged in the early 17th century, since matter in Descartes' physics is defined as "spatial extension" or "Body."

However, it was at this point that alchemical traditions, developed in response to the alchemical traditions imported from China through the Islamic empire (see Lindberg (2007), Needham (1969)),[82] began to display interesting divergences from

---

[81] See Timaeus's speech beginning on 29e.
[82] Lindberg remarks that alchemical tradition probably came from ancient Greece. However, while a natural assumption, this is not corroborated. Needham's work, then, acts as a proper bridge, showing how the advanced matter physics of China especially was filtered into Europe both in specific engineering marvels like the printing press or the horse stirrup or gunpowder, but also in terms of the theory of matter that was built on Mohist and Taoist traditions. The

2.2 Underlier Arguments and Their Types — 75

the orthodoxy on the definition of matter. Namely, three sorts of alchemical interpretation of matter were at play, defined by the debate surrounding the proper methods of physical manipulation and production of alchemical changes in matter. Namely, there were:

(1) *Plenum/Continuum interpretations:* Change in substances must be solely in terms of the form of the substance. I.e., in a shift from water to ice, wetness is replaced with dryness, with no corresponding change in the underlying matter that accepts these qualities/formal features. Substance/matter is therefore the bare subject of form, and it is only differences in form that define differences between subsystems of the world.

(2) *Minima interpretations:* Change in substances is the result of stripping form from matter, and replacing it with a new form. This replacement results in the creation of new core hylomorphic components—new minimal units of material composition (minima)—since each form defines a characteristic minimum component of that form. Substance/matter is therefore the hylomorphic indivisible minimum of composition.[83]

(3) *Atomist interpretations:* Change in substance is the result of rearranging the core components of the substance, i.e., the discrete atomic constituents of various macroscopic substances like lead or gold.

Of these, only the third could produce a consistent account of the method for altering substances in the world. I.e., only (3) had any scientific impact. That is, (3) introduced the now-familiar perturbative method of modern physics as the core means of producing changes of interest to the experimenter. It was this tradition that eventually developed into chemistry, through the work of especially Boyle. Plenum theory saw a few attempts to reclaim explanatory ground, e.g., in the explanation of thermal reactions in terms of phlogiston (Becher and Stahl). However, the success of the method of (3) and the resultant chemistry was far too great. Though atomism would not be "confirmed" until much later, the methods of those assuming something along the lines of atomism produced results.

---

fact that alchemical practice was so prevalent in ancient China, and that alchemical traditions became far more prevalent in Europe around the time of these transfers in technology from China to Europe is taken as reasonable evidence that Chinese alchemy had some influence on the practices in Europe. Needham admits, however, that the precise relationship between Chinese and European alchemy is still unclear. A more thorough investigation is still needed.

83 The minima interpretations saw very little support. They were primarily defended as a philosophical attempt to achieve the best of both the atomist and the continuum interpretations. However, they failed primarily because they could not describe a method for "stripping form from matter," when form and matter were thought to be co-defining.

Importantly, this opened the door for the final rejection of the Aristotelian connection between matters and motions. The characterization of material composition in terms of components and their organizations meant that no longer could we describe particular motions as the result of particular material natures. All material components would need to be able to move in similar ways in order to account for the many composites of the world in the framework of (3).

Having rejected this connection, physicists were able to take seriously the perturbative methods that would prove so useful in especially quantum mechanics and quantum chemistry. No longer was it necessary to describe experimental setups in terms of qualitative changes in the form of the substance. Instead, physicists could describe solely the slight alterations of quantities of motion independent of stuff (substance) and its features. Matter physics then proceeded as a steady precisification of the observable and isolatable motions associated with material systems. In so cataloging these motions, matter physics required sequentially the positing of smaller and smaller components of material systems to carry the relevant motions and dynamics. Each instance of such a posit was eventually rejected in favor of lower level dynamics of smaller things. This brings us roughly to the current day, in which the "most fundamental" material component—the Higgs—is not obviously a thing at all, but is better described as a fundamental self-interaction through which systems acquire their mass.

There is an intuitive sense in which matter physics is about substances. We need only look at examples throughout history to see that the non-rigorous sense of substance—some stuff, usually optically homogeneous—is a core target of investigation. Alchemy is about changing stuff like lead into gold. Chemistry, born from alchemy, is similarly about stuffs like salt and acid and hydrocarbons. Fluid dynamics and solid state physics are both about particular phases of stuff, fluids and solids. Thermodynamics is about the response of stuff to heating and cooling. Early quantum mechanics is about core components of stuff—the specific properties and dynamics of the constituents of the periodic table. Later quantum mechanics is about the measurability and character of stuff in general.

I do not think it unnatural to suppose this. However, a more thorough look at the history, which regrettably I could only sketch here, reveals that this assumption that matter physics is about stuff is far too simplistic. Indeed, most matter physics was developed to explain stuff in terms of more fundamental dynamics. These developments were defined by rejections of things in favor of dynamics, and by the increasing precision of perturbing the stable features of the systems being studied. From the substances of chemistry, to molecules, to atoms, to nuclei, to nucleons, to quarks, to field fluctuations, the progress of matter physics has been defined by a move toward more and more fine-grained dynamics, both in experimental practice and in explanatory modeling.

As I have said, I can only sketch this here. However, Chapters 3, 4, and 5 describe in detail three moments in this progression of matter physics, in the order in which they historically occurred. Together, the chapters compose an inductive base for the claim I have sketched here: that for each thing-posit in history, eventual work in matter physics (or other science) eliminated (or eliminates) this thing-posit in favor of more fundamental and explanatory dynamics. I call this the regression argument in favor of processes.

For the purposes of this chapter, suffice it to say that once again we cannot rule out that matter is processual and not substantial. While the claim that matter is substantial is intuitive, the process realist may claim instead that matter is an umbrella concept meant to express a particular class of interactions. For example, one might say that a process or event is material, or occurs in matter, whenever that process or dynamic event includes as parts gravitational, electromagnetic, and other such physical interactions. In this way, we might say that a new type of matter has been discovered whenever we have discovered an instance in which these interactions are not similarly co-localized. For example, dark matter is a new type of matter present in galactic gravitational dynamics precisely because there are gravitational interactions with no associated co-localized electromagnetic interactions.

2.2.3.2 The Matter as Unifier Argument

Rather than presupposing that matter is substantial, one might instead attempt to argue that matter acts as the unifier of experiments in the world. Essentially, the thing realist argues along lines similar to the robustness or unification arguments above. In every experiment, there is some aspect of the experimental system that is called material. What's more, these same aspects are found unchanged in many different experiments on many different experimental systems. Thus (by virtue of the robustness argument), these material aspects must be unchanging entities to unify the diverse dynamics of those different experiments.

Due to the similarity with the robustness and unification arguments, this argument variant will face similar problems, namely, illicit inductive inferences. In fact, this is little more than a precisification of those arguments. As such, the argument deserves no additional refutation beyond that offered for the robustness argument in particular. The argument is interesting in its own right only insofar as it is the most common robustness or unification argument offered in favor of thing realism. However, this historical commonality can be accounted for as merely the result of the overwhelming focus on substance ontology throughout the history of philosophy, philosophy of science, and the science that developed out of these traditions. If one wishes to describe the world *as composed of substances*, then it is only natural that one will offer primarily arguments for the existence of this matter or that matter, this material feature or that material feature. The process realist, in turn, needs only to note that throughout all of the back and forth about what material things to reify in science, the dynamics within which these things are supposedly located remain untouched by metaphysical controversy. This point will be taken up in more detail in Chapter 5.

2.2.3.3 The Argument from Metaphysical Necessity

This argument rests on the intuition that processes are necessarily twofold entities, requiring both an actor and an act. Essentially, processes cannot be subjectless. Thus, even if some processes can have processual subjects, eventually we must posit a subject that is not processual, or face an infinite regress. In this case, the infinite regress is taken to be an absurdity, and so there must at least be some substantial thing to act as the grounding subject for all less-fundamental processes.

This is a fair argument to make. Of course, the possibility of subjectless processes is not a matter of intuition but of definability. If we can identify processes in the world independent of the identification of some subject for that process, then there is no empirically grounded reason to suppose that all processes must have subjects. We can identify processes in this way, so we cut the problematic regress short. For example, we can identify in quantum field theory transition amplitudes —measures of energetic fluctuations in a system—that are "off-shell." These off-shell fluctuations are by definition fluctuations that are not carried by a real particle. Indeed, the fact that such off-shell fluctuations occur indicates that the idea of particles carrying out these fluctuations is heuristic.

Indeed, this argument is made weak by being so general. The regress is only problematic if indeed we can show it must occur. If all processes were described using subject-laden language (as in "I ran," or "the atom decayed"), then we would indeed be faced with the regress. However, if even one process is not described commonly using subject-laden language, then the burden of proof shifts. E.g., since we commonly say "it is snowing outside" with the implicit understanding that the "it" specifies no real subject, the proponent of the metaphysical necessity argument must provide us reason to suppose that most or some processes are not similarly subjectless.[84]

---

[84] See Sellars (1981) for advocacy for the idea of subjectless processes, drawing on the work of Broad (1959 [1934]) who uses the terms "absolute becoming" to mean becoming that has no subject.

2.2.3.4 The Priority of Stability Argument

A cousin of the metaphysical necessity argument, this argument states change is *defined* in terms of differences in states of affairs. In other words, a change is nothing more than the difference in properties between two states of being. If this definition is accurate, no change can be defined without reference to the static things in which the change manifests. I.e., stabilities, specifically static things, are metaphysically prior to changes and dynamics.

The definition, however, is not accurate. Or rather, it has no ontological import. How we talk about change in natural language constrains, but does not necessitate, a particular ontological model. Moreover, while we often do describe the occurrent dynamics in a system in terms of the boundaries of those dynamics—the initial and final states of the process, as it were—this does not mean we should reify those boundaries. Rather, we should treat those boundaries—the states of affairs and properties therein—as what they are: an epistemic means for defining dynamics in relation to other dynamics. This is explicitly how I argued in Chapter 1, and I will continue to argue this point in Chapters 3, 4, and 5.

More importantly, change is not something that can or should be defined in terms of anything static. This is one area where process realism can draw on a history of explicit arguments against this possibility. Namely, dating at least back to Zeno and Parmenides, arguments exist showing the impossibility of change *provided change is defined in terms of definite properties of things.* In the modern era, philosophers such as Bergson, James, and Whitehead have argued that such a definition is impossible.

The crux of these arguments is as follows. Zeno's paradoxes present us with a tension between an entity that both has a fixed (determinate) property and can change that property. Explicitly, Zeno presents this as a contradiction between an object having a location and the ability to move. If an object has a location, then it is fully described without reference to any other locations. If, however, an object is in motion, it cannot be fully described without reference to at least two locations. So goes the paradox: a property that is changing cannot be fixed, but a property must be fixed to be held by an object.

The key to resolving this paradox has been properly noted by many philosophers of the past. Most recently by Maël (2018).[85] Put simply, Zeno's paradoxes

---

[85] Bathfield also provides some examples of how Zeno's paradoxes appear in modern physics and science as a result of failures to understand this point. See also Silagade (2005) and Atkinson (2006) for discussions of the paradoxes in modern science. See also Papa-Grimaldi (1996) for discussions of how the common mathematical solutions to the paradox miss the point. See also Lynds (2003) for a brief discussion (based on a longer work) of the paradoxes in specific that shows how the

smuggle in a premise that is illicit: that there are static things with static properties in the first place. Since things are static, they will be described and fully understood even when they are not changing. I.e., there is no sense in which a static description of a thing at rest is lacking any information necessary for a complete definition of that entity. In this way, Zeno (and every Parmenidean to follow) sneaks in a metaphysical premise that begs the question: that entities in the world are static.

Notice that it is not simply enough to resist this premise. For example, one cannot simply adopt "causal powers" or "intrinsic states of motion" and still resolve the paradoxes. In order to resolve Zeno's paradoxes, one must accept that there are no such things as static entities. In other words, the resolution of the paradoxes amounts to the negation of the claim that dynamics and change are defined in terms of anything other than further dynamics and change.

The Parmenidean concept of being, which has been so prevalent in European philosophy, is difficult to dislodge. It is ensconced in Plato and Aristotle, and so figures prominently in the Scholastics and Early Modern philosophers who defined the landscape of philosophy for the 20th and now 21st centuries. However, its status as a significant historical influence should not mean that it is immune to criticism. The Parmenidean conception of being has many difficulties in dealing with the concept of change and dynamicity to this day. It fails to account for philosophical positions and success stories from cultures outside of the European traditions. Any argument dependent on the Parmenidean concept of being, such as this and other underlier arguments, will therefore inherit its philosophical problems and contextual quirks.

2.2.3.5 The Common Cause Argument

Another means of arguing that things must underlie processes is to argue that only things can act as common causes. In other words, one assumes that processes are incapable of multiply causing. Thus, when we think there are two events caused by the same preceding event, there must be a thing that explains the evolution of the original event into the two later events.

This argument is little more than another example of the robustness and/or unification argument already discussed. As such, it will face similar challenges. In this case, the assumption that processes cannot be multiply causing is a false premise. Processes are identified in part by their dynamic shape (§1.2). One such shape is a fork, in which one process splits into two. Such a process is a clear ex-

---

paradoxes express a tradeoff between determinate properties and determinable continuous properties (coded language for a tradeoff between substance and process-ontological pictures).

ample of a physical common cause. We have also seen that processes can act as underliers and unifiers of multiple different phenomena. The common cause argument is therefore insufficient by itself.

There is a further problem, related to thing realism, that is worth noting. Namely, things cannot cause except insofar as they are dynamic. In order for a thing to affect change, it itself must achieve its potential to change. In order for a thing to compose and enable an occurrent *event*, it must itself have an occurrent aspect.[86] This means that, at best, things can act as structural, not causal, explanations of the dynamics. This is tantamount to an admission that things do not play a physical-causal role in the dynamics of an experiment, and so will not serve the thing realist very well as they attempt to justify that things must underlie processes.

2.2.3.6 The Symmetry Argument

One of the primary means of justifying the existence of *structures and structural features* in philosophy of science is the appeal to symmetries. The idea is that (a subclass of) structures obtain in virtue of static, quantified mathematical features of a state of a system.[87] These structures have definite relational features—the symmetry relation—that obtain independent of time. Therefore, intuitively, these symmetries cannot be held by inherently temporal entities like processes.[88]

It is difficult to see how this argument is meant to work in detail. The basic argument is simple enough, but symmetries are most often treated as theoretical relational facts that constrain systems and system properties, but are not found in any property of those systems and properties. Another snarl comes from the fact

---

86 See, for instance, Simons and Melia (2000).
87 There are cross-temporal symmetries discussed, especially in the context of quantum field theory where CPT-symmetry is explicitly temporal.
88 Once more, our friends Bueno (2001, 2006) and French (2001, 2006) both offer accounts of symmetries in the literature that are interesting in their differences of ontological commitment. French (2001) is explicit in his commitment to structures qua static features of objects, and takes symmetry arguments in physics especially as a reason to support structural realism (further supported in 2006). Bueno (2001) is (as one might expect given our earlier discussion of these interlocutors) less committed, treating symmetries not as indicators of anything ontic, but rather as an effective tool for drawing inferences about physical systems, especially physical dynamics in the examples from von Neumann and Weyl he presents. See Butterfield (2005) for an account of symmetries that is both less-technical and relatively ontologically neutral. See also Earman (2002), who argues that symmetries are important for establishing *invariances* (coded language for thing-like entities and their properties). See Rosen (2008), who argues that symmetries are actually less interesting in the study and interpretation of physics than asymmetries, and uses this as a justification for a Whiteheadian process interpretation of physics. Finally, see Baker (2010).

that symmetry arguments are most common in quantum field theory, where it is less than obvious that there are entities that can bear the sort of static symmetry relation that the thing realist needs.

I think this argument is meant to express a mathematical-descriptive point: that symmetries are often explanatory and powerful within our theories when they are mathematically defined. For some strange reason, the descriptive power of mathematics has been co-opted by the structural realist literature as their domain alone,[89] and so symmetries are taken to be suggestive of structures and structures alone.

However, processes can manifest symmetries. Dynamic symmetries like CPT-symmetry[90] are examples. Non-dynamic symmetries, like transposition or rotational symmetry, can also be manifest by processes so long as processes can have measurable quantities in small enough spacetime regions. Put another way, there is no reason to suppose that a stable relation—a symmetry relation—entails the existence of anything static. In fact, we can reinterpret all symmetry claims in the literature by noting that processes can not only bear symmetry, they are often the defining features of that symmetry. Symmetries in physics, quantum mechanics and field theory in particular, are defined as the similarity of actions of a system under certain perturbations, where those actions are considered the "reverses" of each other in some sense. For example, temporal symmetry is defined as the similarity of the equations of motion, field action, transition, etc. for a system under a reflection in the direction of time.[91] The symmetry then arises because

---

[89] See Brading and Landry (2006), Bueno (1999, 2000), French (2001, 2003a, 2006, 2011), Psillos (1995, 2001), van Fraassen (2006, 2007, 2008), Votsis (2003, 2005), and Worrall (1989, 2007) for various forms of ontic and epistemic structural realism. Worrall in particular is guilty of the assertion that mathematics is indicative of structuralism, as are all ontic structural realists. An interesting counterexample comes from Earley (2008a, b, c, 2012, 2016) who argues consistently for a "process structural realism" where structures are understood as the emergent order found in systems of multiple processes like chemical reactions or molecular dynamics, similar to Heraclitus's *Logos*.

[90] For the curious reader, CPT-symmetry is a feature of physical laws in quantum field theory first derived by Schwinger (1951). The symmetry is a recognition that field-theoretic transitions exhibit symmetry under simultaneous conjugation of charge, parity transformation, and time reversal. In essence, there is charge, parity, and time-ordering equivalence between transitions that occur in Lorentz invariant fields. So, the universe, or the Lorentz invariant field, will evolve according to the same physical laws if we simultaneously reverse all charge (charge conjugation), reflect all locations of fluctuations through an arbitrary point (parity transformation), and reverse all momenta (time reversal). The symmetry is a dynamic symmetry precisely because it is a symmetry in evolution of systems in time (as opposed to geometric symmetries such as the rotational symmetry of a sphere).

[91] Here, the direction of time is to be understood solely mathematically: we just flip the sign of the temporal variable from positive to negative (or vice versa).

the dynamic evolution of the system is identical whether run forwards or backwards.[92] Thus, there is no reason to suppose that symmetries entail things.

2.2.3.7 The Contingent Thing Argument (Exhibition)

Finally, the thing realist might simply point to a thing. Exhibition is an argument for existence, after all. This is truly the last resort of the thing realist. Yet, it is perhaps the most persuasive argument, simply because it rests on no general assumptions. Rather, the argument is based entirely on peculiarities of particular scientific experiments and models. The Contingent Thing Argument goes as follows. A particular collection of dynamics include reference to an underlying thing, with all of the properties of things: material features and the like. Thus, contingent on this description being accurate, this particular collection of dynamics refers explicitly to a real thing to underlie and undergo those dynamics.

For example, it was once argued that fire is a substance. It had properties that suggested it was a physical thing. Many fires, such as a candle flame, had structures that seemed (relatively) stable. They always had the same color(s). They would always produce the same sensations when touched, and were causally dependent upon the combustion of some other substance, the fuel.

Of course, we later discovered that fire is not a thing. It is no more than a collection of dynamics.[93] Its structure, such as it is, is no more than the balance of interactions between those processes: the motions of air and the combustion products. Its color is no more than the result of a particular process within the system: incandescence. Its dependence on fuel is not actually dependence on a substance, but rather a dependence on a particular sort of chemical interaction: yet more dynamics. In fact, it is no exaggeration to claim that the entirety of this supposed thing and its supposed thing-properties is accurately described in terms of underlying *dynamics*.[94]

So, this example will not work for the thing realist. What about other examples? The issue should be apparent: the very thing that made this argument strong—its contingency—also places a heavy burden on the thing realist. Namely, they must actually produce an instance in the history of science wherein a thing was posited, was successfully retained through all theory changes, was not super-

---

[92] A lot must be said about this, but unfortunately, I lack the tools to enable this discussion at present. The purpose of this work is, in large part, to build the foundation for this future analysis.
[93] C.f. Psillos (1994) for a discussion of this. It should be noted that Psillos approaches the discussion from a structural realist position, and so misses key points about the shift in thinking to thermal *dynamics*.
[94] See, for instance, Bickhard (2009), for arguments to this effect.

seded by a better description of a similar thing, and was not (and cannot be) later described in terms of underlying dynamics as was the candle flame. This is quite the task. No example presents itself at this time. Perhaps, if the thing realist believes strongly enough, an example will appear: a messiah for thing realism.

More importantly, and in all seriousness, as I argue in the next chapter, the thing realist is actually subject to a historically contingent regression argument. Namely, for each thing we posited in the past, science eventually showed how that thing is no more than a collection of dynamics. Given the plethora of examples of this, and their alignment with the general trends of scientific inquiry, we have good reason to suspect that this trend will continue: things will always be but stepping stones to a dynamic understanding of the world.

2.3 General Refutation: From Negation to Position

2.3.1 The Negative: Refuting Underlier Arguments Algorithmically

Underlier arguments all require that somewhere, somehow, we recognize stability in the world. This recognition is of one of two types of stability: (1) an observed stability that is mind-independent, or (2) a stability in our epistemic or referential access to that mind-independent world. This latter type is then divided into linguistic and conceptual subtypes. From this recognition, the thing realist then argues that these stabilities either are or entail static things: structures, substances, etc. This argument that stability entails staticity is either deductive, or it is inductive. Deductive versions rest on assumptions (whether supported or not) about language, about metaphysical priorities, and about the nature of the material. Inductive versions instead attempt to justify the entailment of staticity from stability by appeal to specific cases or collections thereof.

We have already discussed the general refutation of the underlier argument. Namely, it is simply false that stability entails staticity. In order for this to be true, stability would need to be an absolute property of systems. However, stability is relational: something can only be stable with respect to that which is not. The claim that stability entails staticity therefore amounts to the claim that everything that is stable is imperturbable. I.e., the thing realist must infer from the existence of something that is unchanged to the existence of something unchanging.

Recognizing the failure of the thing realist's assumption—that stability entails staticity—is the key to refuting every underlier argument that has been, or will be offered. One might even construct an algorithm out of these refutations. First, locate the offending premise. Second, articulate how the particular stability of interest can be represented by a comparison of two dynamics with different characteristic time or energy scales. Finally, if possible, show how this dynamic representation is superior at capturing the practice of science in the context of the particular system in which was found the stability to begin with. This last step is merely icing on the cake: all the process realist needs to do in order to refute an underlier argument is to demonstrate parity between dynamic and static representations of stability. The fact that dynamic representations are generally superior only acts as further, but unnecessary, reason to endorse pure process realism.

Crucially, the thing realist can offer no inductive support for the claim that stability entails staticity, even in particular instances. We discussed this in the context of two of the stronger underlier arguments: the robustness and unification arguments. There, we noted that in order to offer inductive support for the stability entails staticity premise, the thing realist would need to have access to some material fact about the world and its constituents that would justify the inductive comparison between every possible dynamic system. I.e., when we inductively infer the existence of a thing through underlier arguments, we assume that not only is the entity in question stable in many systems under many interventions and dynamics, but that it is stable in every feasible system, under every feasible intervention. This sort of inductive-risk-laden universality is, perhaps, acceptable to the devout thing realist, but no one can call it epistemically modest.

Moreover, even granting that there are thing underliers to act as robust unifiers of experiments and their descriptions, these underliers are accessible and useful to us only insofar as they are dynamic. For example, when we (supposedly) use electrons to probe the electromagnetic properties of some system, the thing realist would have us believe that this entails the existence of electrons qua things. However, what we are actually using to probe the system is the electromagnetic interaction processes. "Electron" is just the word we give for this interaction. It is not the probe which we are using, but rather the prob*ing*: the interactions and propagations we already know enough to make use of.[95]

This lack of inductive support for the offending premise of the general underlier argument means that the thing realist can only justify their claims with deductive principles. This in turn means that their support for the existence of things can only come from principles that are at least as strong as the claim that stability entails staticity. Moreover, these deductive principles cannot come from experience: it is impossible to experience a thing because experience requires interaction, i.e., dynamics. The deductive principles must be a priori, and pure of inductive support.

---

[95] C.f. Hacking's "use" argument for entity realism (Hacking 1984). Importantly, I have merely co-opted Hacking's argument by applying it unaltered save for the rejection of the implicit underlier argument Hacking adopts.

Thus, underlier arguments for things are inevitably question-begging: they must at least assume that which they are attempting to prove.

In contrast, the process realist can offer arguments based solely in experience, or from induction. This was the point of the continuity argument from Chapter 1, and of the various refutations of thing-underlier arguments above. This breaks the parity between process and thing realisms. Most underlier arguments are defeated once we recognize that they do not act as arguments for thing-underliers over process underliers. However, the continuity argument, and the various sample arguments offered through this chapter in favor of particular process underliers suggests that, whenever we are justified in inferring to underliers, we will only be justified in inferring to process underliers.

2.3.2 The Positive: Explaining Stability

Something stronger can be said. If underlier arguments all fail because they cannot justify that stability entails staticity, this indicates that stability is not the sort of thing that can even be explained by staticity, independent of any entailment relationship. In fact, this is the case: stability is an inherently dynamic notion, as I argue in this section.

Every claim about stability is comparative. Consider: "the table has a stable shape," or "the molecule is stable," and "the microtransaction business practices of predatory game publishers stably produce revenue." One of the unifying features of these claims qua stability is that they are meaningless without a comparison class. I.e., "the table has a stable shape" is equivalent in logical content to "the table has a shape" if no comparison is specified or implicit. E.g., in a logically possible world composed only of a single table, the claims are identically made true or false. It is only when we specify some time scale over which we are considering the table—a time scale over which other changes *do* occur—that we notice that the table's shape does not change in that same time scale. We might similarly specify an energy scale for comparison: the table has a stable shape *within my house* because my house contains no highly energetic processes or systems. Put the table in the center of the sun, and it will not retain a stable shape. Finally, we might simply compare the table to other systems that are less stable. E.g., the table has a stable shape compared to the shape of a collection of gaseous molecules.

However, no matter how we specify the comparison class of the stable entity, we will be appealing to processual notions. I.e., we will always explain the difference between the stable entity and its comparison in virtue of the changes each undergo either independently or when subjected to the same perturbations. E.g., this Uranium-216 atom is stable and that Uranium-239 atom is not precisely because when both are allowed to decay naturally, Uranium-216 will decay in milliseconds, while the Uranium-239 will decay in nanoseconds. The processes of decay (of the transition to a lower energy state within the nuclear-energy-shells of the respective nuclei) occur over different time scales.

Every claim about stability must therefore be made in reference to a comparison class of dynamics. In addition to this, stability claims come attached to dynamic contexts. Consider: "The microtransaction business practices of predatory game publishers stably produce revenue." The claim is still comparative; what would count as unstable production of revenue? The claim is also obviously context sensitive. The production of revenue in various industries occurs over different time scales. For example, the production and selling of grand pianos can take years, and involves a very small market and so the revenue stream from this industry is indexed to very long time scales with large bursts of income periodic to those time scales. In contrast, the production of glassware occurs over days at most, and has a relatively constant consumer base. The revenue stream in this case occurs far more frequently and in smaller spikes.

What we mean by stability in this instance is clearly relative to more than just a comparison between specific processes. Rather, we must also define what sort of stability we are interested in with respect to other, non-comparable processes. Is the size of the influxes of revenue or is the time between them more important for determining the stability of the revenue stream? The answer is that it depends on what our context is. For a company, long gaps between revenue influxes may be acceptable or even preferable so long as the influxes occur frequently enough and are large enough that the company can engage in its spending processes—paying its workers, investing in more production capacity, paying off loans, hoarding money in offshore accounts, etc. For an individual looking to hand-make these products, other factors will be more important—affording food, paying for medical coverage, etc. Thus, while a company may say that the revenue from making and selling pianos is stable, a family will not necessarily say the same *in virtue of the other economic processes of interest to each respectively.*

Thus, stability has (at least) two conceptual features:
(1) Stability is comparative between processes of change.
(2) Stability is relative to a context of other processes of change.

These features of stability are explicit and suggestive in physics. When we experiment on a physical system, those aspects that are stable are said to be so in virtue of having a characteristic binding energy greater than the perturbing energies of the experimental interventions. Claims about stability are made true both by the comparison between systems subjected to the same perturbations *and* by the recognition of which perturbations we are actually performing on the system.

In short, stability is like "faster than": it can only exist as a comparison, and only as a comparison between temporally extended entities. As such, stability requires the existence of dynamics and change in order to be an operational concept. In other words, processes are ontologically prior to stabilities; change always supersedes rest.

This is a point that is worth stressing. Much of the history of philosophy, and of philosophy of science, has been centered on explaining change—in terms of differences of states of affairs, differences in properties, causal connections, causal powers, potentialities, etc.—but this work assumes a metaphysical priority relation that is illicit. Namely, it assumes that change is the thing to be explained, and that statics are brute. However, this simply does not accord with experience. This priority relation should be reversed: it is change that is brute and primary. Stability can only exist in a world defined by brute change.

Corollary to this, it may be worthwhile for the GPT that I have subscribed to to adopt an additional feature as process-identifying in order to account for stability claims. That is, something along the lines of "characteristic energy or time scale" could explicitly ground claims of stability in physical sciences to allow for clear hierarchies between the experimental processes of interest and the processes that persist through experimental intervention. While the GPT does not need any additional machinery to account for linguistic and conceptual data, I would advocate that we need this additional classification/differentiation of processes in order to account for the practices of physics. This becomes especially important if we wish to construct an account of spatiotemporal locality using the GPT. However, this new classification (relativity to a characteristic energy of destructibility/perturbation) may be possible to construct as a constellation of other features/classifications of processes.

2.4 Conclusion

One of the common refrains against process metaphysics is that it is revisionist. The implication is that in order to justify or motivate a move to a process ontology, and a process interpretation of science or language, we need sufficient reason to give up our existing ontology of things. The weight and accident of history then oppresses such a move. After all, things were sufficient throughout history for interpreting and understanding science and language in the majority of cases. What could process ontology add to this?

Of course, this is similar to the mistake made by those who initially rejected quantum mechanics for contradicting classical intuitions. Just because thing-talk has succeeded in the past does not mean that thing-talk has more success than process-talk. What's more, in this case, it is only the history of Western philosophy and science (and arguably not even the science) that suggests any success of thing-talk. Both a broader scope and a narrower focus reveal that thing-talk was not nearly as successful as the anti-revisionist would have us believe, and there were indeed plenty of other philosophical approaches that emphasized processes over things. In fact, Chinese, Indian, Japanese, and Korean historical philosophy all enjoyed great cultural and scientific successes while centering on ontologies of process. The idea that process ontology is a revision is therefore not an absolute judgment, but is dependent on a particular cultural context and the resulting linguistic and scientific data we choose to admit as evidence for an ontology; process ontology does not revise our ontology, it revises orthodox (Christian) European (mostly English, French, and Italian) ontologies.[96]

Nevertheless, it is worthwhile to point out those linguistic, conceptual, or scientific considerations that can only be handled with a process ontology. This chapter produces one such. Namely, stability cannot be understood with things alone, while processes are clearly capable of explaining stability. In particular, there is no such thing as absolute stability, and even if there were, too many of the stabilities we describe in scientific models are relative stabilities. In fact, stability is conceptually empty without dynamics: the whole idea is that there is a system that is unchanged in response to various perturbative forces. Moreover, we often make explicit claims about relative stability in our scientific models.

Consider, as a final example, the photoelectric effect. In the relevant experimental system, we have essentially an undriven capacitor formed of a plate of metal and a separated wire lead, with both connected by an unclosed circuit. Photons incident on the plate will cause a current to pass through the capacitor system (i.e., to flow from plate to wire lead) when they have some minimum frequency. The modeled explanation for this[97] is that electrons in the plate are dislodged from the plate to jump to the wire lead only when the photons have the appropriate energy, which requires that they have some minimum frequency. I.e., the elec-

---

[96] Needham spent his entire academic career arguing against our assumptions that this Orthodoxy is somehow scientifically superior to others, in particular those from China, India, and Japan. The key point to remember is that Chinese medicine, physics, and engineering was in many ways far more advanced than European equivalents for most of history, as measured by their relative standard of living and life expectancy and the specifics of their advances. What's more, many of the great advances in especially chemistry and engineering in Europe can be traced to previous work done in China that were later transmitted through the Islamic empire into the hands of scientists like Boyle.

[97] See Einstein's (1905a) paper containing the light-quantum argument.

trons in the plate are stably contained within the plate up to a certain energy of perturbation. I.e., the electrons are only ever relatively stably bound in the plate.

We should take this seriously. Not only do we have no need of things to explain and describe experimental systems, we have reason to believe that processes are the explanatory entities of all aspects of our experimental systems, including the underlying persistent or stable aspects of our experiments. It would not be an exaggeration to say that the practice of science requires this. In short, processes, and processes alone, underlie experimental dynamics.

Interlude: Two Shifts in Method

We have so far discussed process realism in a relatively abstract context. Arguments of the earlier chapters were mostly about what it is possible to infer from experiment in general, rather than from specific experiments or historical moments. In the coming chapters, we shift our method, instead seeking to evidence process realism explicitly in the actual practice and history of physics.

This will mostly take the form of three arguments about macroscopic thermal systems, microscopic thermal systems, and nuclear systems. In each case, there are specific arguments being made, but there is also a structural argument implicit in the background of these chapters. Namely, by presenting the chapters in this sequence, and noting the parallel to actual historical developments in physics, I begin to make the further *inductive* argument that for every thing-posit in physics, eventual work has shown or will show that the thing is *nothing more than a collection of more fundamental dynamics*. This inductive argument is also an implicit regress: in the actual history of physics, things have been posited, shown to be collections of dynamics in smaller things, and then these things have been analyzed in exactly the same manner.

The regressive part of this induction presents us with something like a parity between things and processes (note that the inductive part does not present such a parity). In the historical regress, processes are posited to explain things, and then things to explain processes. Implicitly breaking the parity of this regress is that field theory has no natural particle interpretation (Malament 1996) (nor even a substantial field interpretation, see Baker 2009). As I have already written above, the Higgs field, and mass, should be understood as a subjectless process of self-interacting (like "snowing"). Further breaking this apparent parity is the epistemic modesty of process realism. As Chapter 1 argued, processes are necessary posits for understanding experiments and experimental practice. As Chapter 2 showed, things are nowhere near necessary posits, no matter how desirable they may be.

We will also explicitly break the parity in Chapter 5. In this chapter, I will show that thing-interpretations of certain systems—namely, nuclear systems—are impossible because they are explicitly contradictory. Process interpretations, in contrast, remain consistent and plausible. Thus, parity will be broken in a third manner.

The inductive regress argument, then, will be left as an implication of the chapters to come. However, it is important to frame these chapters in light of this inductive regress. In this manner, we see a third, *inductive* method for arguing for process realism, equally general as the arguments of Chapters 1 and 2.

There is one additional methodological point to make as we move from Chapters 1 and 2 to 3, 4, and 5. That is, we will be placing a much greater emphasis on the explanatory power of process-posits within scientific theories. This should be contrasted with the more general ontological sufficiency of processes that we saw in Chapters 1 and 2. In Chapters 3, 4, and 5, it matters most that the processual parts of various theories, models, and experiments that we infer from our descriptions of experimental systems are acting as explanans for the explananda within those experimental systems. That is, these later chapters involve arguments to the effect that processes are the only necessary posits for our explanations to make sense, not that they are necessary for our descriptions or inter-model inferences (qua inferential method in analytic ontology).

However, these two views of processes—as ontological ground for descriptions/linguistic cohesion or as explanatory posits—overlap in an important way. Namely, I will assume in what follows that the referent of a theoretical term or description is explanatory just when that referent is a necessary precondition for the explanatory inferences we draw. For example, if I explain the blueness of the sky as the result of scattering of light from the sun through our atmosphere, all and only those entities that are necessary preconditions for our inference from "the sky is blue" to "there is scattering of light in our atmosphere" count as explanatory entities. In many cases, these explanatory entities will be identical to those that allow for us to infer the truth of various statements in general language. However, we must always remember that our primary goal in science is not necessarily truth (or "capital T truth"), but rather is explanation. This goal of explanation constrains our inferences, practices, and language in ways different from the constraints of speaking the truth, or of accurately describing the world. Where necessary, I will point out these differences in the chapters to come.

Chapter 3
The Candle Flame: A Process-Realist Analysis

3.1 Introduction

In the last two chapters, I argued first that we must commit to processes—which were categorized according to the GPT as general, subjectless, contextually individuated, determinable, and measurable dynamic entities akin to activities—within our experiments. Second, I argued that inferences to any further entities within our experiments (and especially things like substances, atemporal structures, objects, etc.) either fail or fail to act as inferences to something other than processes. These arguments taken together support that things cannot appear as ontological posits in our theories, so long as we expect our theories to be primarily interested in describing and explaining possible or actual experiments.

However, the thing realist has one final bastion: to rely on the explanatory power of things. Namely, the thing realist argues that the explanations of scientific models would be impossible, or else a miracle, if things were not real.[98] Perhaps, therefore, we might commit to things as explanatory entities without reifying them. I.e., we hold firmly to a fundamental-process ontology, but allow for things to appear as real-enough entities in our epistemology.

I argue in this chapter that this fails. As many others have already noted, the no-miracles argument is fallacious,[99] it adopts false premises,[100] and even modest versions of it fail.[101] Moreover, there are many instances in scientific theories where the explanations offered seem not to involve things at all.[102] I tend to

---

[98] C.f. Putnam (1975, 73) for the first explicit example of this "no-miracles" argument. The argument has taken many forms in the years since Putnam's paper, and appears in spiritual ancestors of Putnam as well. See, for instance, Barnes (2002), J. Brown (1982), Boyd (1989), Busch (2008), Dellsén (2016), Frost-Arnold (2010), Lipton (1994), Lyons (2003), and Psillos (1999, ch. 4). Note that of these, only Psillos and Putnam explicitly argue for a thing-categorical explanans. The others may only be construed as offering such arguments, given the targets of their arguments (i.e., the antirealists who offer pessimistic meta-inductions).
[99] See, for instance Howson (2000, ch. 3), Lipton (2004, 196–198), Magnus and Callender (2004). See also Menke (2014) for a criticism of the no-miracles argument based on a different probabilistic framing in terms of likelihoods, see Sober (2015, 912–915).
[100] See van Fraassen (1980) and Wray (2007, 2010) for arguments that the success of science, or the fact of its successful explanations, require no explanation themselves.
[101] E.g., the version in which the no-miracles argument is called an abduction, or inference to the best explanation.
[102] See the contributions in Bueno, Chen, and Fagan (2019), and in Eastman and Keeton (2008).

agree with this analysis. However, in this chapter, I will argue from a different angle. Namely, I argue that:
(1) Things will inevitably be explained by processes, and
(2) Things explain only insofar as they have associated, explanatory processes.

Together, these points spell the explanatory defeat of things within scientific theory. Notice, however, that my argument does not force us into an antirealist position. Rather, my argument further strengthens the claims of Chapters 1 and 2 by showing that processes and only processes are explanatory in our models. In other words, the processes of our models and theories are doing all of the explanatory work. Things are just an unfortunate byproduct in science of the historically influential Parmenidean concept of being, especially the belief that processes metaphysically require things as subjects.

To argue for (1) and (2), I analyze the example of the candle flame, as it was understood both pre- and post-chemical revolution. This example exhibits two interesting features:
(1) Scientists and philosophers of the past thought the candle flame (and fire generally) was thing-like (i.e., substantial).
(2) Scientists of today describe fire as the rapid cascading oxidation of the fuel in an exothermic process of combustion, i.e., they describe it as a process.

We are therefore led to ask the historical question: how did scientific inquiry refute the former intuition and lead us to our current one? Importantly, scientific inquiry did not reveal that any of the reasons for which historical figures adopted position (1) were wrong. Rather, scientific inquiry resulted in the rejection of underlying assumptions made by those who adopted position (1). Namely, I show that science revealed the mistaken assumption made by proponents of (1) of the static character of the various features of the candle flame, replacing them with explanatory and descriptive dynamics instead.

To begin, I will evidence the thing realist's posit in history, briefly presenting the history of phlogiston and its refutation (§3.2). I will then present an analysis of the candle flame system, making use of historical descriptions of each of its thing-like elements. In this analysis, I show that the key explanatory move is to eliminate the thing-ness of the candle flame in favor of underlying dynamics (§3.3). After showing that every explanatory task in this system can (and was historically) performed using processual, not thing, posits, I will turn to the question of what role things *could* still play in these explanations. Finding that these roles can equally be played (and will be played in future history) by processes, I conclude that there is no explanatory role for things to play (§3.4). I conclude by noting that this extended example of the candle flame, analyzing the explanatory tasks and successes of chemico-thermal models, mirrors perfectly the arguments I presented in Chapters 1 and 2 (§3.5). This example therefore acts as a historical instantiation of those general arguments.

3.2 The Flame and Phlogiston, the Thing Realist's Posit

At face value, the candle flame appears to be a substantial entity as much as any other mesoscopic object we interact with daily. When we poke it, we feel something. It has persistence and characteristic shape. There are clear properties (all sorts!) we can ascribe to it, such as "hot," "yellow," "7 cm long," "persistent throughout dinnertime," and "capable of igniting this piece of paper." While it may undergo changes, e.g., its characteristic flicker when blown upon, the candle flame appears to persist in much the same way as a table or chair: some properties remain unchanged, while others change. For most intents and purposes, the candle flame is as much a thing as our solid, work-a-day tea kettles and coffee mugs.[103]

It should be no surprise, then, that the candle flame (and fire in general) was treated as a substantial entity (i.e., things or quantities of stuff) for centuries. Beginning in the Aristotelian tradition in which fire is one of five material causes,[104] fire would continue to be treated as a substantial thing through the Scholastic period and a significant portion of early modern period. Scholastic thinkers explained many of the experiences of fire in terms of primitive substantial properties of this substance. Most famously, Johann-Joachim Becher and Georg Ernst Stahl in the 17th century proposed that fire (and combustion) is the manifestation of a substance called phlogiston.[105]

---

103 Note that the candle flame doesn't meet the common sense idea of solidity, or non-interpenetrability of everyday things. Neither does it meet the common sense idea of transportability.
104 It is sometimes suggested that Heraclitus was the first to posit fire as the matter of the world. However, I believe this attribution to be incorrect because (a) it falsely considers Heraclitus to be one of the material monists when what little we know of him suggests that he was quite adamantly opposed to this orthodoxy, and (b) the fragment used to attribute this view to Heraclitus (the world is an ever-living fire ...) can be read as emphasizing the metaphorical comparison to the flickering of fire rather than the material constitution of fire. The fragment in question stresses that the world comes to be and is annihilated in equal proportion, suggesting that it is this change and flow in the world that makes it comparable to fire.
105 See Becher (1669, and later 1708, *True Theory of Medicine*) and Stahl (1700). Stahl was the one to coin the term "phlogiston," replacing Becher's term "Terra pinguis" from 1669. Both are held responsible for the actual physical claim about the substantial (material) character of fire.

Today, we tend to think of these views as the result of bad science, or of an illicit use of dreaded metaphysics.[106] However, one finds that these positions are based on prima facie empirical arguments. Namely, they are based on the observation of special sorts of features in the world. We might summarize these views as follows: the candle flame (and fire in general) is a thing because it has a collection of features indicative of thing-ness. These are features like:

- *Spatial shape:* The candle flame has a spatial shape that is definite at a moment in time.
- *Spatial property/relational structure:* The flame has various regions defined by definite properties like color, location, or the chemical potentials in those regions.
- *Countable properties:* The candle flame is one flame, it has three color/heat regions.
- *Particularity:* The candle flame is here and not there (both spatially and temporally), and cannot be there if it is here.[107]
- *Continuance:* The candle flame is a continuant, in that it is definitely identified at moments of time and continues through moments (at any time, I can point to the candle flame and identify it definitely and completely). I.e., it is not (or not obviously) occurrent.[108]
- *Material:* The flame has the properties we associate with other states of matter (causal potency for material systems, physical extension, perturbability, etc.).
- *Materially perturbable:* We can see, touch, and hear the flame.
- *Causal, motive, potent:* The flame affects other material systems.[109]

Each of these features bears similarity to the features of other, paradigmatic things like rocks. This is further confirmed by comparing these features to the features of processes (§1.2), which shows that the candle flame seems to have none of the identifying features of a process. Purely spatial shape and structure, for example, are typifying of things, not processes (processes do not in general have a characteristic

---

106 See Laudan (1981), Poincaré (1952 [1905]), Putnam (1978).
107 Particularity can also be couched in terms of persistence, being a locus of change, countability (being one "particular instance" of its kind), non-instantiable, independent, discrete, unified, or simple. These different categories are drawn from Aristotle's accounts of "Ousia" (commonly translated as substance or subject). See Aristotle, *Metaphysics* 1017b16, 1038b35, 1041a4, 1042a34, *Physics* 200b33, *Categories* 2a13, 3b33.
108 C.f., Johnson (1921).
109 Note again that this list does not include transportability, nor does it include solidity, which are some of the features that differentiate substantial things from substantial stuff like phlogiston.

shape, "square" for instance). Therefore, to most, it would seem natural to call the candle flame a thing.

However, this presents us with a difficulty. The candle flame still bears these thing-like features. Candle flames still have shape, we can still see them, we can still touch them, they still have internal structure and bear external relations. They are still (apparently) particular, continuant, and singlet (countable) systems. It is not as if the candle flame stopped being hot or teardrop-shaped with the advent of the chemical revolution. So, why is it that we no longer think of the candle flame as a thing? This question generalizes. Namely, if the features that once led philosophers and scientists to consider the candle flame to be a thing cannot prove that the flame is in fact a thing, then do we have any reason to suppose that there are things at all?

It would be easy to dismiss this as but another example supporting the pessimistic meta-induction in the history of science, as indeed is done in Laudan (1981).[110] Historically, the thing-ness of the candle flame (e.g., phlogiston, or the element "fire") was a theoretical posit that was later proven to be untenable and superseded by a more accurate theory. Thus, this example fits the pattern of this meta-induction perfectly. One might therefore take this as evidence that antirealism is the only acceptable and epistemically modest interpretation of this particular science.

This is too quick. Scientists of the time did not resort to antirealism. Never did they claim that the features of the candle flame—the "thing features"—were not actual or real. Nor did they reject that their models of the candle flame were describing one or more real entities. Rather, they merely redefined and explained the thing-features. For example, Antoine Lavoisier's work between 1770 and 1790 on oxidation and reduction of iron, tin, and sulfur was taken as direct evidence that combustion reactions (i.e., oxidation reactions) did not involve the release or stimulation of some "combustible substance," as had been claimed by Becher

---

[110] See also P. Lewis (2001), Psillos (1996), and Saatsi (2005) for reconstructions of the pessimistic meta-induction. Saatsi's work is especially interesting, since it is a defense of the meta-induction. See also Saatsi (2015, 2016). Saatsi (2016) seems to be an explicit opponent of process accounts (e.g., "dynamical systems theory") in philosophy of science, criticizing Lyon and Colyvan (2007) for their "explanatory indispensability argument" for dynamical systems theory. I should note here that the dynamical systems theory, while process-realist-adjacent, has some key differences to my own pure process realism, most notably the lack of explicit consideration of the history of process ontology and its literature. Dynamical systems theory, construed as a mere mathematical tool lacking ontological import, does provide some useful theoretical machinery for understanding dynamic shape, however, and so may be useful for the process realist in some way.

and Stahl.[111] Instead, combustion was described as an interaction between an organic substance (or metal) and what Lavoisier termed "oxygene."[112] Lavoisier himself described his work in 1783 as evidence that phlogiston was imaginary.[113] Other properties of the fire and flame received a similar analysis with the development of both the new chemistry of the late 18th century and the development of optics and electromagnetism. Both the radiation of light and heat from the candle flame came to be described not as some intrinsic aspect of a substantial element of the world, as had previously been claimed by the scholastics. Rather, these radiative properties were explained as the result of emission of light and energy, two deeper underlying substances the action of which we could directly measure with our eyes and implements like a thermometer respectively. Crucially, the substances were left unexplained until much later, but scientists were comfortable using the *action* of these substances—the processes—both as explanans and explananda of the phenomena.

This collective work to analyze the underlying mechanistic cause and properties of fire and flame led, eventually, to the development of a complete redescription of the substantial being of the candle flame. In 1848, Michael Faraday collected these analyses into a lecture series dedicated to this very redescription called "The Chemical History of the Candle Flame." A series of six lectures, Faraday explicitly states that the goal of his analysis is a full causal description of the candle flame:

> We come here to be Philosophers ... whenever a result happens, especially if it be new, you should say, "What is the cause? Why does it occur?" and you will in the course of time find out the reason ... (Faraday 2016 [1848], Lecture 1)

Moreover, this causal description is reductive: Faraday describes each manifest material feature of the candle flame—every property that could serve as evidence of the substantial being of the candle flame—in terms of underlying interactions, mo-

---

[111] It is perhaps more accurate to take Lavoisier's work as a refutation of the Aristotelian argument for substantial fire, rather than as a refutation of Becher and Stahl.
[112] The discovery of oxygen, however, was not Lavoisier's alone, as it was built on isolation experiments done by Henry Cavendish (2011 [1766]) and Joseph Priestley (1774).
[113] See McKie (1935) for a biographical account of Lavoisier's work, and Lavoisier (1770–1790). An interesting historical quirk is that, when more detailed scientific study reveals a previously accepted substance to be physically no more than a collection of processes, terms like "imaginary" and "illusionary" are employed. We will see this same language appearing in descriptions of the atom, the nucleus, and the particle as we progress. Perhaps this terminology is meant to evoke the Platonic distinction between the true and the imaginary world, or perhaps the terminology is meant as a rebuke against believers in the substance. I will not be analyzing the historical significance of this language in great enough detail to judge.

tions, and other processes. In other words, the goal of such a model is to fully understand what it is to be the candle flame, its shape, its generative cause, and its matter, and all of these are described processually, as I show in the next section.

In short, it would be incorrect to claim that phlogiston is an example of bad science. Rather, it was bad metaphysics. Phlogiston and elemental fire theories sought a static explanation for what were essentially dynamics of the candle flame system. Thus, when dynamic explanations and definitions of these features were offered, the phlogiston and elemental fire theories were superseded. The mistake was not to posit that there were such properties in the world, nor to posit that such properties were indicative of some real underlying entity. The mistake was to posit that this underlying entity was thing-like and not processual. Exactly how we are meant to understand these supposed material features of the candle flame is the subject of the next section.

3.3 Explaining Things Away

3.3.1 What Do We See in the Flames?

We begin our analysis with a short primer on everything going on in the candle.[114] Roughly, there are three explananda of interest in the candle:
(1) The production of the bright yellow, teardrop-shaped flame (its causal origin)
(2) The maintenance of this flame (how it persists and by what means)
(3) The features of the environment that constrain and allow the production and maintenance of the flame

By explaining these explananda, we will also explain the other properties of the flame—its color regions, its various internal energetic structures, etc.—simply in virtue of achieving fine-grained analysis. Presumably, the production of the flame, along with how and by what means it persists, will together explain any feature of the candle flame we see. Of these, (1) is the simple process of combustion: where and under what conditions combustion occurs. (2) is the cycle of activities that produces the stability of the flame: the combination of convection currents to bring new air to the combustion region, the melting of wax and capillary action to bring new fuel to the combustion region, and steady heat radiation to initiate com-

---

[114] It should be noted that this is an abridged and paraphrased description of Faraday's explanation in his lecture series. The analysis I offer is anachronistic for the sake of clarity to the modern reader, but I will show throughout that my explanation is a reasonable paraphrase and deviates only slightly from Faraday's.

bustion. (3) is the combination of goings on outside of (1) and (2) that must occur so as not to disturb (and in extreme cases, destroy) the production or maintenance of the candle flame. (3) includes factors like the specifics of the shape of the candle (including its production and maintenance), the kinds of convection that must occur, the composition of the wax and the air surrounding it, etc. Together, these three features form a complete explanation of the candle flame hypothetically sitting before us.

A candle flame is produced initially by bringing oxygen, paraffin (in the typical wax candle), and energy together at the top of the wick of a candle. More accurately, one first brings a match near the candle wick. The match releases highly energized particles (primarily light) which impart their motion to the paraffin molecules of the wax. These paraffin molecules, now moving faster, break free of the binding forces which hold them in the solid candle, forming liquid paraffin. This liquid is then transported up the candle wick by capillary action, at which place the paraffin is further energized into vaporous form. There, the now-vaporous paraffin and oxygen energized by the radiation from the match collide. These collisions result in the breaking of the chemical bonds of the paraffin and oxygen, and the resultant scattering and loss of energy of the scattering products allows the recombination of these products into carbon dioxide and water vapor. These products then emit light as they release energy in their bonding and through their incandescence. The energy from this collision, and the consumption of oxygen from the air causes air to flow upward and around the combustion at the wick. Thus, the products of combustion incandere as they flow upward, forming the bright yellow teardrop shape we call the candle flame. This is how the candle flame is produced, feature (1) above.

This flame is then maintained by the radiation of heat from combustion. As combustion occurs at the base of the candle flame, the resultant emission of energized particles (heat radiation) performs the same activity as the match did in the genesis of the candle flame. Paraffin melts, then is transported up the wick by capillary action, then vaporizes, then combusts when in contact with air. More interestingly, the radiation from the combustion process facilitates the continued flow of oxygen up and around the candle base to the combustion zone. As energized particles are emitted from the combustion zone, they energize the surrounding air, causing it to flow upward. This flow of air upward creates a pressure differential in air below the candle, causing it, too, to rise. The convection currents that result from this sequence bring a steady supply of oxygen to the combustion zone and maintain the shape of the candle flame. This completes our discussion of how the candle flame is maintained, feature (2) above.

Lastly, we are concerned with what environmental factors must obtain for the production and maintenance of the candle flame. These are many; depending on how specific one wishes to make these environmental factors, one may obtain a list of factors of arbitrary length.[115] However, there are two environmental factors that are of particular historical interest. First is the shape of the candle's wax base. In order for a stable candle flame to persist, the convection currents must be symmetric so as not to produce any oscillation or fluctuation in the flow of the air and incandescent combustion products. This means that the wax must be symmetric. In addition, the wax base must be such that melted wax does not simply run down the sides; the wax base must form some sort of vessel for the liquid paraffin, and the vessel must be undisturbed by continued heat emission from the combustion zone. This is achieved by having a wax base that is both symmetric—so that no segment of the edge of the wax is more exposed to heat than another—and wide enough that the heat differential from the combustion zone provides more heat to the solid paraffin close to the wick. Convection currents then cool the edge of the wax base, while the wax at the center of the base is melted into liquid paraffin. It is in this way that (in Faraday's words) a "beautiful cup" is formed that contains the liquid paraffin, preventing the symmetry of the wax base from being disturbed by run-off liquid paraffin or uneven heating.

The second environmental factor of interest is the consistent supply of oxygen. In order for combustion to occur continuously, thereby maintaining a persistent and stable candle flame, all of the interactions that produce combustion must be capable of continuous occurrence. I.e., oxygen must collide with vaporized paraffin in the combustion zone continuously. In order for this to occur, both paraffin and oxygen must be continuously transported to the combustion zone, at which place they can interact to combust. The paraffin is supplied by the candle by way of the melting and capillary action processes already discussed. The oxygen is supplied by the environment by way of the convection currents surrounding the candle flame.

Notice that the environmental features appear as specific constraints on the processes that occur in the production and maintenance of the candle flame. We may very well remove these features entirely from our explanations with no loss by simply incorporating these constraints into the descriptions of the descriptions used to explain production and maintenance. For example, we know that the

---

115 We might, for example, say that "the environment should not have a limited amount of oxygen" is a feature that must obtain for a persistent candle flame. We may then specify every way that an environment may have a limited amount of oxygen: being contained in a bell jar, being in a bubble in the ocean, being fed oxygen from a tank in space, etc. Or, we could group these cases into types: being in a closed system, being in an oxygen-free environment that is temporarily supplied oxygen, etc. In this way, we obtain lists of different cardinality, even though they are not interestingly different.

base of the wax candle must be symmetric in order for the flame to maintain a stable shape. However, it is only because the radiation from the combustion region and the convection currents that surround the combustion region are circularly symmetric that the wax cup is formed symmetrically in the first place. Rather than playing an active role in the maintenance of the candle flame's shape, the constraint of the wax cup's symmetry acts only as a constraint on the *continued* symmetric flow of air around the candle flame. This additional environmental factor—the symmetry of the wax cup—is irrelevant; it is only the symmetric flow of the convection currents (and symmetric melting of the wax) that makes any difference. In other words, the reference to a symmetric wax cup acts as a placeholder for reference to the symmetry of the processes of convective air flow and radiative melting.

3.3.2 What Do We Explain about the Flames?

The brief explanation of the candle flame in §3.3.1 is already suggestive. Both the genesis of the candle and its persistence are explained in terms of triggering processes, cycles of processes, and balances thereof. The environmental factors—what initially appear to be structures and static properties of things—play a role in the explanations only insofar as they evoke the balance of dynamic activities. We seem to have explained the candle's broad nature in terms of processes and a few underlying things like oxygen and paraffin.

However, the analysis goes even deeper than this. In fact, we can show how the broad processual explanation of section 3.3.1 translates directly into specific explanations of every feature of the candle flame system. In short, anything that *could* act as an indicator of a static thing is generated, maintained, and learned about via processes, and so is explained dynamically. This includes, most notably, material features of the supposed thing underliers: oxygen, paraffin, and the like.

To see this, we must first separate out the relevant explananda of the candle flame. It is here that we will draw on Faraday's analysis explicitly. The first step in Faraday's analysis is to isolate a collection of features of the candle flame. These features serve as the basis for further, causal analysis in the course of the explanations Faraday presents. These properties are many, but roughly fall into three categories: (1) the form of the candle flame, its shape, size and other monadic properties of, and relations between, its parts; (2) the matter of the candle flame, the "actual composition" of the candle system; and (3) the productive origin of the candle flame, how it is created and how it persists. Specific members of these three categories are listed in Table 1 below. We shall proceed through each category independently in the next three subsections. More attention is given to the two types of property that one might reasonably suppose are not dynamic: formal and material. However, it is worth working through an example of a productive property to set the stage for later discussion. Moreover, we will focus on single examples of properties from Table 1 so as to limit the amount of exegesis in favor of more philosophical discussion.

Table 1: A list of features of the candle flame that Faraday seeks to explain, grouped into three categories

Type of Property	Formal	Material	Productive/Motive
Token Properties to be Explained	– The shape of the candle flame – The shape of the wax – The shapes of the three regions of the flame (blue, gray, yellow) – The relative positions of the three regions of the flame – The color of the flame in its three regions – The heat variance of the three regions of the flame	– The chemical phase/state of the wax as it is burned – The chemical phase/state of the wax that is transported up the wick as fuel – The chemical phase/state of each of the three regions of the flame – The chemical phase/state of the pre-combustion materials – The chemical phase/state of the post-combustion materials	– The nature of the transport of fuel from the base of the wick to the candle – The nature of the production of more fuel by the flame (i.e., the persistence of the flame) – The nature of the production of emitted light – The nature of the production of emitted heat – The nature of the production of post-combustion materials from pre-combustion materials

3.3.3 Explaining What Needs Explaining

3.3.3.1 Explaining Productive Features

Let us consider as an example the movement of fuel from the base of the candle to the combustion region. Faraday remarks to his listeners:

> Notice that when the flame runs down the wick to the wax, it gets extinguished, but it goes on burning in the part above. Now, I have no doubt you will ask, how is it that the wax, which will not burn of itself, gets up to the top of the wick, where it will burn? How is it that this solid gets there, it not being a fluid? ... *Capillary action* conveys the fuel to the part where combustion goes on, and it is deposited there, not in a careless way, but very beautifully in

the very midst of the center of action which takes place around it. (Faraday 2016 [1848], Lecture 1)[116]

Faraday points out to his readers that there is a processual gap that must be explained. Namely, we see, in the first place through the emission of light from the candle flame,[117] that combustion is occurring in a location spatially removed from the fuel of the combustion. Put another way, the activity that we directly perceive in the candle—the propagation of light from the flame to our eyes—is not obviously spatially or temporally continuous with the activity that creates the fuel for combustion—the melting of the wax. We must therefore explain how fuel is moved from the wax base of the candle to the location at which combustion occurs. We must also explain how oxygen, a necessary component of combustion, is consistently delivered to the area of the flame in which combustion occurs: the blue portion of the candle flame in a ring around the wick and the gray region. Finally, we must explain how combustion results in the perceived effects of both blue and yellow light radiation. In short, we must describe how the initial process of melting wax is carried through to the final process that we see directly.

Faraday describes this processual sequence in his lectures in roughly the order in which they occur in an actual candle. First, the heat from the ignition mechanism (e.g., a match) melts the wax by heat radiation. Next, capillary action delivers the melted wax to the top of the wick. At the top of the wick, the heat of the ignition mechanism vaporizes the wax, allowing it to flow out from the wick to the edges of the candle flame. At the edges, the vaporized wax comes into contact with oxygen that is consistently delivered to the contact area by convection currents, which are in turn caused by the heating of surrounding air by the candle

---

116 As a point of interest, notice how Faraday describes this process of capillary action as "beautiful." Interestingly, Faraday refers to many aspects of the candle and the candle flame as beautiful, but the aspects which are beautiful are always processes of formation, of balanced interactions which preserve the stability of the candle, of production, etc. Beauty is most clearly associated with process when Faraday states at the beginning of his lectures: "I hope you will now see that the perfection of a processes—that is, its utility—is the better point of beautify about [the candle]. It is not the best *looking* thing, but the best acting thing, which is the most advantageous to us" (Faraday 2016 [1848], Lecture 1). This trend of describing beauty in terms of processes, not static qualities, is an interesting historical throughline in the work of scientists after the scientific and chemical revolutions.

117 We later discover the exact relative location of combustion through more precise interventions, as we shall discuss later in this section.

flame.[118] The contact between oxygen and highly energetic vaporized wax initiates the combustion process, wherein the paraffin vapor ($C_{25}H_{52}$) and the oxygen (O_2) are converted into water (H_2O), carbon (C), carbon dioxide (CO_2), and a great deal of heat. The C and CO_2 is energized enough to be incandescent, releasing blue light. The heat radiating from this combustion provides the initiating process for the melting of more wax and the flow of additional convection currents. The convection currents then move the water vapor and still-incandescent carbon dioxide up into the yellow part of the candle flame. Having been moved out of the hottest part of the candle flame, the carbon dioxide emits yellow instead of blue light.

Thus, the explanation of the processual gap between lighting the candle and seeing light rests on the following sequence of processes:
(1) Emission of heat from the ignition source
(2) Melting of the wax
(3) Capillary action
(4) Vaporization of the melted wax
(5) Combustion with oxygen (emission of blue light and heat)
(6) Convection current initiation
(7) Flow of incandescent combustion products up with the convection currents (emission of yellow light)

Starred processes (combustion and initiation of convection currents) then act as a new set of ignition processes,[119] beginning anew the sequence depicted above by melting the wax and bringing new oxygen into the combustion region. Our processual sequence therefore fills the gap between ignition and observation of the candle, explaining how one event leads to the other, *and* explains why the candle persists in its activity.

This style of explanation will occur throughout our later discussion of formal and material properties. First, we notice a processual gap between some activity in the candle system and a dynamic interaction between the candle system and our own senses (i.e., a process we engage in or respond to directly). Second, we fill this gap with additional dynamic transitions, interactions, and motions. In so doing, we

---

118 Note also that incomplete combustion takes place nearer to the wick, in the gray region of the flame. This is because there is characteristic diffusion of oxygen transfer into this region, resulting in less-than-sufficient oxygen to enable complete combustion of the evaporated paraffin.

119 Here we should note that there are no new processes here except as arbitrarily differentiated. In order to understand the dynamic balance of the system, we speak about these processes as if they are in a particular sequence. However, they roughly begin and end simultaneously, and so the construction of sequences does not constitute an identification of different (countable) processes.

construct an explanation of the processes we see involving a single, continuous sequence of dynamics: one process after another from start to finish.

Note that the dynamic shape involved in our explanations will not always be a pure sequence. I.e., we do not need to show that there is exactly one process following and preceding each other process in order to successfully explain any part of the candle system. Rather, we will see loops, cycles, forks, intersections, and all sorts of dynamic shapes appear in our explanations, especially for more complicated systems and features. Indeed, the existence of loop- or cycle-shaped dynamics is a key identifying feature of the processes underlying structural features of our systems, as I show in the next section.

Each of the processes that fills the gap is indeed a general subjectless process, i.e., an occurrence in its own right that exhibits the features of processes described in Chapter 1. We can show this simply by listing the features of processes with which we identify them, and showing that the processes in the candle flame have those features. Taking capillary action for an example:[120]

(1) *Generality:* Capillary action is not a particular. It cannot be located in a definite spatiotemporal point or region, although it can be localized in a region of spacetime by various means.
(2) *Subjectlessness:* Capillary action is an occurrence in its own right, not a modification of some other entity. We can describe it as the process by which paraffin is transported to the combustion region, but it is not thereby a modification of this or that paraffin (or similarly of this or that wax). To test this, we simply note that capillary action is not identified by its occurrence in a particular wick or involving a particular liquid to transport, even if certain measurable specifics of the action are dynamically contextualized to specific interactions with particular wicks and fuels (see below).
(3) *Occurrent nature:* Capillary action is a temporally extended occurrence. No moment of time can define/identify capillary action, its effects, its causes, or its features.
(4) *Measurability, not countability:* Capillary action is not unitary (we do not count capillary actions in the world, nor would we say that one system has fewer capillary actions than another). Capillary action is measurable by various means, e.g., the flux of gravitational potential energy or the flux of pressure differentials (we would say that there is less capillary action in one system than in another).

---

[120] Feature (7), according to which processes are not changes, but can be measured or partitioned into changes, is omitted. This feature is a corollary of (1), (2) and (4) in the scientific context.

(5) *Determinability:* Capillary action is not determinate, but can be determined through appropriate measurement practices. E.g., we can isolate capillary action from other processes, and determine its energetic features, by means of various experiments (see below).
(6) *Contextual individuation:* We identify capillary action processes in part by their dynamic context. I.e., this capillary action in a symmetric candle is not that one in an asymmetric candle because of the difference between the dynamic heating of the wax, and the differences in the resultant dynamics of combustion and convection that alter the pressure differentials in the wick and so change the rate and effect of capillary action.

Each of these features is relatively simple to exhibit. However, of particular interest are features (2) and (6), subjectlessness and contextuality. Of the features of processes, these are the two that least obviously identify capillary action and the other processes in the candle flame. The key to doing so is to note that our means of identifying these processes in the candle flame follows the same prescription as we saw in Chapter 1. Namely, we do not identify these processes by the presence (or absence) of some subject, but rather through a collection of interventions, perturbations, and successive observations of the isolated system. I.e., we identify these processes by their dynamic context.

Our knowledge of these processes—both those which define the processual gap and the processes we use to fill the gap—comes from our ability to interact with the system to isolate them. For example, we know that capillary action brings the melted wax to the top of the wick because we can take melted wax, put it in a beaker, introduce a long wick leading far away from the beaker (and the source of heat), and, after waiting a suitable period of time, light the now transported wax at the other end of the wick. Faraday also describes other situations that act similarly to the wick and the melted wax. In particular, he describes an experiment wherein he puts a colored fluid in one beaker, places a wick in the beaker leading to an empty beaker, and then waits as the empty beaker is slowly filled with the colored fluid. One can even see the moment-to-moment action of this process because the fluid is colored. Finally, Faraday notes that replacing the wax or the wick with some material that will not undergo capillary action (e.g., anachronistically, a metallic wick or a heavily polarized liquid) prevents the formation of the candle flame.

The combination of these three experiments allows us to know that the process of capillary action is occurring. We isolate *where* the process is occurring by recognizing that the system is analogous to the beaker filled with liquid and the string to the empty beaker. We then note that the analogous situation of the beaker reveals *that* the fluid is transported. We can then perform interventions on this beaker system to stop or inhibit the transport process such as changing the wick or altering the fluid.[121] This, in turn, reveals how we might intervene on the melted wax and wick system to test its capillary action. Upon performing these interventions, we discover that inhibiting the process of capillary action prevents the initiation or continuation of combustion in the candle flame.

Crucially, we inhibit the process of capillary action by altering the dynamics in the candle flame that enable it. Namely, the melting of the wax, the interaction between the wick and the fuel, and the combustion processes going on at the top of the wick. We notice that merely altering the wick or altering the fuel is not sufficient to change the capillary action process. It is only when the alteration of one or both *would result in a different interaction between wick and fuel* that the capillary action is inhibited. In particular, if the wick and fuel have insufficient adhesion interactions, or if the fluid fails to have sufficient cohesion self-interaction, the capillary action will fail to occur. This reveals to us *how* the fluid is transported. Thus, we know where, when, how, and that capillary action occurs to bring fuel from the candle to the top of the wick, providing us with an understanding of the next step in the dynamic sequence from melting to illumination.

Thus, the process of capillary action is identified by its dynamic context, *not* by some subject or object. This generalizes. We are led, in the case of productive features of the candle system, to identify these productive features with specific processes or sequences of processes. In our example, we identify the production of light in the candle with the sequence that begins with the melting of wax and proceeds through capillary action, combustion, and incandescent radiation.

This is, perhaps, not very surprising. Production is processual, after all. We therefore might expect explanations of production to be processual. However, we will see that the style of explanations given for productive features of the candle apply generally to formal and material features of the candle as well.

3.3.3.2 Explaining Formal/Structural Features

Once we have admitted that the various productive features of the candle flame are processes and are explained by processes, we are forced to admit that the candle flame system *with all its features* is inherently processual. That is, since the

---

[121] Faraday's demonstrations involve replacing the wick with salt and the melted wax with alcohol. This alteration allows for a better demonstration of the capillary action. Then, he changes the salt to a different material, and notes that the capillary action no longer occurs. As we will later discover, the explanation for capillary action is in terms of the adhesion forces between the fluid and the capillary. In short, it is processes of electromagnetic attraction, and the imbalance between the fluid-fluid attraction and fluid-capillary attraction, that explains the process of capillary action itself.

candle flame is defined in part by its occurrence (its processual, temporally extended existence), continuant features of the candle flame like its structure will not exist independent of context. In fact, they will depend on the occurrent features of the candle flame system. When the processes of the candle flame are occurring, the candle flame will have a structure and material character. When those processes are absent, there will be no candle flame, let alone a structure or a material character of the candle flame. We therefore expect that the explanations of structural and material features of the candle flame will be made in terms of processes, and the same processes that we saw in §3.3.3.1.

Moreover, since the structures and material features of the candle flame will be identified and explained by the same processes as those we saw in §3.3.3.1, we should expect that these structures and material features will exhibit all of the identifying features of processes. They will inherit these processual features from the processes on which they depend. I.e., the structure of the candle flame is explained by and dependent on a dynamic context, specific underlying dynamics that make it countable, determinable, general, and subjectless, and is itself necessarily temporally extended, unidentifiable at a moment in time. Put simply, structures are stabilities, not staticities. They are therefore necessarily dependent on and defined by processes. As I will show, the structures in the candle flame do indeed have this processual character.[122]

Let us consider the shape of the candle. That is, we will discuss two properties: the shape of the wax base and the shape of the flame. Both of these are investigated by Faraday in the opening lecture, and share many of the same features. Faraday begins his analysis with the shape of the flame and the shape of the wax—what Faraday calls the "beautiful cup."[123] As we have already briefly discussed, this cup and the shape of the flame itself are formed and maintained by the air currents which flow around the candle flame as it burns. As the candle flame emits heat, it excites the air surrounding the flame, causing it to rise. Concurrently, the emitted heat melts the wax at the base of the wick. The excited air, in rising, causes cooler air to rush up and around the candle flame to fill the void left by the heated air. This rising cooler air—the convection currents—keeps a continuous heat sink next to the edges of the wax. This means that the heat which melts the wax at the base of the wick is quickly dispersed into the convection currents near the edges of the candle. Therefore, not enough heat is delivered to the wax at the edge of the candle to melt it, with progressively more of the wax receiving

---

[122] Note that my analysis will mirror work done on molecular structures by Earley (2008a, b, c, 2012, 2016).
[123] See Lecture 1. Faraday's full explanation for the shape of the candle flame and wax cup are also found in Lecture 1. I reproduce simplified versions of these explanations here.

enough heat to melt the closer to the wick one measures. Hence, a cup of wax is formed, with melted wax held within.

The rising convection currents also supply a steady stream of cool air to be heated by the candle flame. These convection currents flow up and around the flame, directing the products of combustion up and out of the flame's incandescent region. The source of heat being the total combustion taking place at the candle's wick, air is heated by a greater degree closer to the tip of the wick. This ensures that as the air flows around the candle flame, it will flow around the central axis of the candle flame, and will converge to this axis as it rises. This means that the products of combustion, including those that incandesce to form the characteristic yellow region of the flame, will similarly flow symmetrically around, and converge to the central axis of the candle. This symmetric flow of air and incandescent combustion products produces exactly the characteristic teardrop shape of the candle flame.

Importantly, the causal relationships between these convection currents and the shapes of the candle and the wax cup are established by Faraday through a series of interventions. These interventions allow him to isolate the specific interactions and processes which give rise to certain observable features of the candle system. We know that it is the convection currents which cause the candle flame to take on its teardrop shape because disturbing these convection currents disturbs the flame's shape. This disturbance can be introduced by either inhibiting the flow of the convection currents, e.g., by blowing on the candle flame or altering the aerodynamic interaction between the wax and the air currents. Faraday remarks that this is why "novelty" candles that lack the standard symmetry of a cylindrical candle fail to form a consistent shape: such candles inhibit the flow of convection currents around the flame through their irregular aerodynamics. Similarly, Faraday remarks that tipping the candle reveals the importance of the convection currents as well. The more one tips the candle, the more the convection currents are inhibited. A small degree of tipping leaves the candle flame roughly undisturbed since the convection currents are uninhibited. In contrast, the wax cup forms asymmetrically even with a small degree of tipping as the convection currents cool the edge of the wax in the direction tipping more efficiently, causing the edge of that side of the wax to be much higher and fully formed than the opposite side. With a large degree of tipping, the candle flame begins to sputter, no longer holding a consistent shape as the convection currents are no longer able to flow consistently around the flame. Similarly, a large degree of tipping destroys the wax cup entirely.

These interventions reveal the causal relationship between the convection currents and the shapes of the wax and the candle flame. I.e., disturbing the convection currents in specific ways predictably disturbs these formal properties of the candle. However, we must notice two points in this analysis. First, that the interventions are actions taken by the observer: tipping the candle, blowing on the flame, introducing additional inhibiting material to the side of the wax cup. Moreover, the interventions are revealing of some causal relationship between what is intervened on and the system of interest because they produce some novel, noticeable change in behavior of the system. We notice that the candle flame no longer maintains a stable shape when we disturb the convection currents, instead sputtering and flickering.[124]

Second, we notice that the interventions are on specific activities of and surrounding the candle. In this case, the convection currents are being disturbed so that they can no longer flow in the appropriate way. We intervene not on the composition of the air, but only on the *motion* of the air and the interactions that guide this motion. Thus, not only is the intervention itself a process carried out by the observer, but that which is intervened upon is a process as well.

What we learn from these two points is that, contrary to the thing-realist intuition of a static shape-property inhering in the candle flame, the shape of a candle flame just is the processes that produce and maintain the stability of the candle flame. I.e., it is through a balance of processes—in this case, heat radiation and air convection—that the property of "teardrop shapedness" comes into being and persists. We learn about and explain this property of shape through dynamic interventions. Our interventions, activities that we perform, trigger some new dynamic event in the system, and we infer from this to the existence of dynamic interactions in the system that we have disturbed. After performing several interventions,

---

[124] A few interlocutors have remarked to me that one might think of this intervention not as a process at all, but as a simple alteration of the state of the system. I.e., there is no active change that is playing any role, but rather only a passive difference. I take it that this remark trades on the idea that we do not care how we intervene on the system, only that we have successfully altered it. It is then, according to my interlocutors, this alteration, and not some process, that acts as the salient feature in our explanations and acquisition of knowledge about the system. While I understand the desire for such an account—it is motivated by the reasonable equivocation of mathematical variables and aspects of a system—I find this account to be mistaken. It would be feasible if we could observe a system in two perfectly defined states without any dynamic interaction between us and the system, but we cannot. Moreover, it is precisely because we observe the system changing that we are inclined to infer anything in the first place. Consider: it would not be sufficient for any inference about the system to merely observe that it is in some state X. Rather, we need to see that the system *comes to be* in state X from some other state. In short, it matters how a system comes to be in X. This is just another way of saying that what guides our explanations are active interventions in which we bring about a change in a system, not some mere fact of difference. (Think of an example of a system in two states successively that is brought there by two different means—this shows that the process matters: it is the difference maker). Thanks to Gal Ben-Porath and Jim Woodward for this comment.

on the system in this way, we are able to further infer that these dynamic interactions in the system balance in some appropriate way so as to produce some form of stability in the system: shape in this example.

This inferential process is exactly what generalizes to all formal properties of the candle flame, and to other systems besides. We have the following general scheme:

(1) We intervene (Process 1, *intervention*) on a system, and observe the change in activity of the system (Process 2, *observation*). E.g., we blow on a candle and observe that the emission of light to our eye alters.
(2) We infer from Process 2 that Process 1 caused Process 2 by some series of dynamic interactions. E.g., our blowing caused the change in the pattern of light emissions.
(3) If we have performed the appropriate intervention or series of interventions, we are then able to infer the dynamics that lead from our intervention to the observation (Process 3). E.g., the flow of air around the candle is what defines the region of motion of the incandescent combustion products. Thus, by disturbing the first process (the air flow), we dynamically produce the second, that constitutes the change we observe.
(4) We repeat this process until we can chart some collection of *mutually interacting* processes within a system that, when balanced, produce a steady cycle of activity in the system (Process 4). This steady cycle then defines one or more stable aspects of the system, and so we infer that the system possesses a stable formal property in virtue of this cycle. E.g., the cycle of heating and cooling that continually guides incandescent combustion products up in a teardrop shape.

Formal properties and structures are identified with these balanced, cyclic processes in a system. This means that structures are defined by a dynamic shape (cyclic) with a dynamic context (the processes which feed into and flow out of the cycle). Moreover, the processes within the cycle are defined in the same manner as we have seen in previous sections. Namely, they have all of the identifying features of processes (generality, subjectlessness, measurability, determinability, contextuality, and occurrent nature), and we determine these features by means of our dynamic interventions on the system.

We should not be surprised that we arrived at this result. Formal properties are meant to be stabilities. However, we only understand stabilities when we know how they can be perturbed, to what degree, and in what manner. In other words, the stability must be continuously maintained, and contextually defined, as per our discussion in Chapter 2. The stability only has meaning insofar as it can respond to its dynamic context. We will see something similar in the next section, on material features of the candle flame.

3.3.3.3 Explaining Material Features

Material features are perhaps the most difficult to assess, primarily because there is a hidden distinction between two different types of explanation associated with two senses of "material feature." On the one hand, we might say that the material features of the candle flame are those that indicate that the system is material. I.e., we look for the identifying properties of matter in the system, and then seek to explain these. On the other hand, we might say that the material features of the candle flame are the compositional features of the system at a moment in time. I.e., we look for a reduction of states of the system to some collection of entities or properties in the right configuration or with the right character. Let us label these two sorts of explanation:

(*Identifying Material Features*): Those features of the system that identify the system as material

(*Compositional Material Features*): The features of the system's composition, whenever this composition is a configuration of material subsystems

We can separate each of the features listed in Table 1 (§3.3.2) into two separate, related features. For example, we could seek to explain the chemical phase/state of the wax in the combustion region insofar as it identifies a structural feature of the system. In this case, we might say that at each moment of time, there is a property called "material phase" of the system defined and identified entirely at that moment for each region of the system. Explaining this would involve explaining how and that we identify that the system attains this property of material phase.

Alternatively, we could seek to explain the chemical phase/state of the wax in the combustion region insofar as this composition is indeed a composition. At this point, we are no longer explaining material properties of the candle flame, but rather explaining the mereological nature of the candle flame. In order to establish that this mereological nature entails that it is material, we would need to defend the further claim that the components of the chemical phase of the wax in the combustion region are material in character.

In both cases, what counts as material will play a major role in what and how we explain. Unfortunately, there is no clear definition of material we can make use of here. However, we need not answer this question here. Instead, we can simply note that, if we wish for "material" to mean "substantial or thing-like," the properties and entities that we identify as material properties and entities had better be thing-like. I.e., for the sake of argument, we can replace the above explananda with:

(*Identifying Material-Thing Features*): Those features of the system that identify it as a thing. I.e., the features that are definite, are discrete, render the system countable, grant the system a definite location in space and time (make it particular), and

(*Compositional Material-Thing Features*): Those features of the system's composition, whenever this composition is a configuration of things (i.e., identified as above)

The upshot is this: if the features we identify in manifestly material systems are processual and not thing-like, we need only reject that being material has anything to do with being thing-like. We will not be required to revise our claims about material and non-material systems (the same distinctions will hold as normal between, e.g., bodies and souls).

Unfortunately for the thing realist, our analysis of the candle flame results in exactly this conclusion. We begin with the identifying material features. Here, we notice that Faraday implicitly rejects that the candle flame can be identified as a thing at all, simply in virtue of redescribing the candle flame as a system of dynamics. The contrast with Becher and Stahl is here made apparent, since "fire" is no longer associated with a particular substance in which inhere the standard determinate, discrete, definitely located properties of things. In other words, Faraday assumed that the flame is not itself a thing, possessing its own material features, but is at best composed of things: molecules and (anachronistically) photons and the like.

Take, for example, the kinesthetic material features of the candle flame: that when we touch the flame, it burns us (it is causally potent to other material systems). Implicit in Faraday's explanations of this phenomenon is that the flame burns us because of some underlying cause. I.e., the flame contains interacting paraffin and oxygen that, in interacting, produce a great deal of thermal energy. This thermal energy is translated up through the incandescent region of the flame by convection currents that carry the energized combustion products. Thus, when an unsuspecting observer places their finger within this region, the energized combustion particles interact with that finger, translating their energy into the finger, resulting in a sensation expressed vociferously as "ouch, that's hot." Put simply, the candle flame does not possess any material features itself. What might be taken to be material features of the flame itself are no more than processes of (supposed) material constituents.

We can therefore quickly reject the project of explaining any identifying material-thing features in the system; we will not find any if the system is not a thing.

We must now, instead ask, what are the compositional material-thing features of the system. I.e., can we claim that at any moment in time, in some definite location, with determinate identifiable features, the candle flame as a system is composed of thing-like entities?

Answering this question is the subject of the next section. Briefly, our answer will rely on first answering the question of what explanatory role mereological claims can have in explanations of features of the candle flame system. I argue that, insofar as a mereological claim about the candle flame system involves things, it cannot be explanatory.

3.4 What Explanatory Role, Things?

3.4.1 New Thing Terms Appear in Our Explanations …

So far, we have seen that most of the explanations of the candle flame and its features are processual. They involve, depend on, and are about processes. Nevertheless, the thing realist will argue that, because the features of the flame itself are explained by the motions and interactions of further things (its material composition), these material features still serve as positive evidence for thing realism. I.e., if things appear in the explanations of the features of the candle flame, then this suggests that the candle flame is composed of things. If it is composed of things, then even if it is not itself a thing, things are still present in the world and play an explanatory role. So goes the thing-realist argument.

For now, we must assume that the thing realist is still interested in reifying things within the candle flame system in virtue of the study of that system. That is to say, the thing realist is not (in this argument) reifying things within the candle flame system in virtue of an analysis of those things in particular. We have no evidence (yet) that these material constituents are things themselves. Rather, for now, we only assume that they exist as vehicles for the relevant processes within the candle flame. In other words, the thing realist's argument is that, even if all of the explanations of the candle flame system are processual, and all of the features of the flame itself are defined by said processes, analysis of the candle flame system nevertheless results in the reification of at least one thing as an underlier or subject for those processes.[125]

---

[125] In the next chapter, we will analyze the arguments for the particular thing-ness of the compositional entities in the candle flame, i.e., molecules and atoms.

We are therefore led to ask two questions. First, do new things actually appear in our explanations of the candle flame system? Second, if they appear, what role do they play in our explanations? In answering this second question in particular, we are looking for some reason to suppose that things exist on the basis of our examination of the candle flame system alone. In other words, at this point in the regression, it would be illicit to appeal to an independent investigation of the supposed material constituents of the candle flame system in order to reify them. We can only ask: in examining and intervening upon the candle flame, is there some feature or resultant observation that necessitates (or even suggests) the existence of things for the relevant explanations.

3.4.2 ... But Things Play No Role in Our Explanations

The simple answer to these questions is: things play no role in our explanations. While new thing terms do appear in our explanations and descriptions of the candle flame system, they do not appear qua things. They are mere placeholders: words that indicate to us what sorts of interventions we have performed or would need to perform in order to identify the relevant, explanatory dynamics.

Let us consider more closely the chemical state and phase of the pre- and post-combustion materials. As we have discussed briefly above, combustion occurs in the blue region of the flame in a concentric ring around the wick. For paraffin-based candles, this ring is the region where vaporous (energized) paraffin ($C_{25}H_{52}$) and vaporous oxygen (O_2) collide and produce incandescent vaporous carbon dioxide (CO_2) and water vapor (H_2O). Thus, we know both what materials are present before and after combustion, and how these materials manifest (their chemical phase).

We now wish to explain three things: (1) how these material features come to be (and be present) in the combustion region, (2) what role these material features play in the processes that we have already used to define the formal properties of the candle, and (3) how we come to learn of these material features of the candle system. In our analysis, we find the following:

(1) The material features come to be (and be present) in the combustion region because of temporally prior processes.
(2) The material features are important only as (a) placeholders for further processes, and (b) identifiers of process types; e.g., "vaporous" is an important feature of the paraffin because it tells us that the paraffin interacts more energetically and with more dispersion, and "water" is important in the post-combustion convection processes because it identifies that this is indeed a hy-

drocarbon combustion process (as opposed to a more general oxidation process).
(3) We learn about these features by, again, intervening on the system and observing the dynamic event that is triggered. Unlike the formal properties, we do not look for cycles of processes in this case, but rather for same-kind interactions across many interventions (I.e., we do multiple different things and observe similar effects).

For the sake of brevity, I will consider only one of the material features: the phase and existence of water vapor after combustion.

Faraday considers this feature of the candle system in Lectures 2 and 3, Lecture 2 being where he shows that water is produced in the combustion reaction and Lecture 3 where he shows how water comes to be produced through combustion and where the necessary materials come to be in the combustion region of the candle flame. First, Faraday holds a cooled and dried spoon above the candle flame and notices that it becomes dim with condensation. Since this test can be performed in a dry-air environment, the condensation indicates that there is some process in the candle flame which brings about the dimming on the spoon. Before we can investigate this process further, we first need to know the character of the dimming, i.e., we need to know how it is that light comes to be less intensely reflected off of the spoon. We suspect, having seen similar dimming effects in the condensation of water on cool surfaces that the spoon is reflecting light because water vapor is rising from the candle flame and condensing on the spoon. Hence, Faraday first turns the spoon over and places potassium in it, knowing that water reacts strongly with potassium. Observing that this condensation does, in fact, react strongly with the potassium, we conclude that the condensation is probably water because it reacts like water. As a further test, we run an electrical current through the condensation and then burn the products. Since the products burn with the pale blue flame of hydrogen combustion, and since we can perform this combustion in an evacuated, closed system, we have further evidence that the original material which we electrolyzed had the chemical composition of water, hydrogen and oxygen. We therefore conclude that the condensation is, in fact, water.

Since water is neither contained in the wax of the candle nor in the surrounding air,[126] the water must be produced in the combustion process. This then entails

---

[126] Implicitly, we assume that we have set up the experiment in dry air, which means that we have already tested that the surrounding air does not contain, or contains a minimal amount of water vapor.

that the combustion process is a special kind of oxidation reaction, namely hydrocarbon oxidation. In other words, the presence of water allows us to infer that the process at the heart of the candle flame is a specific interaction between hydrogen, carbon, and oxygen. We now have a complete dynamical explanation for the presence of water at the locus of our intervention (the spoon). Water vapor is produced through the interaction of hydrogen, carbon, and oxygen in the blue region of the candle flame, where hydrogen and carbon come to covalently bond. The water vapor remains energized, inheriting the excess energy of the energized paraffin and oxygen, and rises as vapor following the convection currents surrounding the flame. The water vapor then interacts with the cool spoon, distributing the energy of its motion (its heat) into the heat sink of the spoon. The water vapor begins to move more slowly, and thus condenses into liquid.

There are three points to note in this explanation. First, while it may seem obvious, the water vapor comes to be present in the candle system dynamically through the combustion reaction. I.e., the candle system acquires the material feature of "containing water vapor" dynamically. In general, this is true of all material features of the candle system. The material makeup of the candle system is dynamically acquired at all stages, and this is reflected in our explanations. This means that we explain the material state of the post-combustion flame *in terms of processes*; we do not explain the processes of the flame in terms of material states.

Second, just as with the formal properties of the candle system, we learn of this material feature (and others) via a series of dynamical inferences from specific, targeted, interventions. We interact with the system to disturb the process of water vapor rising by introducing the cool spoon. We disturb the process of water moving as vapor (i.e., moving with characteristic dispersion and freedom) by cooling it with the spoon, which slows the water's motion and causes it to condense. This intervention—moving a heat sink above to the candle flame—impacts the activities of the system—vaporous motion—such that we can isolate a specific activity that was present before our intervention—the vaporous motion of water. This isolation allows us to infer that the vaporous motion of the water must have been triggered by the production of water in the candle flame, allowing us to identify the causal connection between the combustion process and the process of vaporous motion.

This leads us to an interesting related point. That is, while we cannot explain the processes in the candle system in terms of states, we can make use of states in order to both identify processes and identify their connections to each other. In other words, we know that the combustion in the candle is a hydrocarbon combustion process because water is produced and rises from the candle flame as vapor. This is not the same as saying that the combustion process is a hydrocarbon combustion process because water is produced. We can say we know what the process is because of our detection of a specific state or material feature, but this does not entail an ontic claim about the nature of the process. Notice: we still learn of the production of water through a series of interventions and inferences about dynamics. We therefore infer the state from the processes.

The third point to notice is that the material nature of "the water" serves three purposes in our experimentation on and subsequent explanation of the candle system:
(1) "Water" acts as a unifying feature of the two interventions we perform and the processes we are interested in, namely, the volatile reaction with potassium, the condensation process, the vaporous motion process, the electrolysis process, and the combustion process.
(2) "Water" acts as a means of labeling an isolated (or isolatable) sub-process within the candle flame.
(3) "Water" acts as a means of defining the causal connection of various sub-processes.

We will consider each of these in turn.

The first role that "water" plays in our experimentation and explanation is to act as a unifying identifier of the processes in the candle flame. I.e., it is because we already assume that water *both* reacts strongly with potassium and condenses on cool surfaces that we are able to make descriptive claims about the causal connection between the combustion in the candle flame and the dimming of the spoon. I.e., this is what lets us say that this is a hydrocarbon combustion process: all such processes produce water.

However, we are only interested in this production of water for the effects that it has on various interventions we perform, e.g., the effect of condensation, or the effect of a reaction with potassium. Insofar as we are interested in the material features of the candle system, we have no interest in whether water is itself a thing. We care only that our interventions reveal the specific character of the combustion process in the candle flame and the convective motions of the air, and that these processes can be perturbed and detected through well-defined acts of intervention.

Moreover, the unificatory role of "water" requires no assumption that water (whether the conglomerate or the individual chemical molecule) is a thing at all. In order for water to be an identifier of combustion, we require only that there is some feature of water-containing systems that is uniquely and consistently identifiable, and that this feature is stable under the interventions we are performing. Allow me to elaborate. In order for "water" to act as a unifying feature of the dynamics in the candle flame, we need to be able to identify its presence. If we could not, we would never posit it in the first place. In order to identify it, we must be able to intervene on various systems in such a way that at least one feature of all of these systems is revealed and comparable to the features revealed in the other systems. To use a toy example, if we take liquid from a still pond and liquid from a river, put both in the same clear container, and drop a stick in the container, we observe the same refraction of light from the stick in each case. The refraction of light through the liquid, or (in more thing-realist terms) the refractive index of the liquid, is similar in both systems under the same interventions. Thus, we say that both systems similarly contain water, the thing that obtains this refractive index or performs this refraction.

However, we need more than similarity in order for "water" to act as a true unifier of the disparate systems. In particular, we need at least that the feature we observe as stable—refraction in our toy example—is stable *under the interventions we perform.* If we intervened on the two containers of water by, e.g., heating them until they completely evaporated, the refractive index would not be a viable unifier of the two systems. In this case, we would refer to some other property of the water (e.g., its boiling point) to unify the two systems.

Returning to our candle flame example, the unification of vaporous motion, combustion, condensation, potassium-reaction, and electrolysis provided by "water" is dependent on our ability to identify in each process (each subsystem of the candle) a feature that is relatively stable under the interventions performed on the candle system. Such a feature could indeed be the thing-ness of water, or its essential properties. However, it need not be. In this case, the two-process sequence of electrolysis and closed-container combustion serves adequately as the unifying feature. I.e., at every stage of the processes of the candle flame, and in every intervention process involving "water," we can consistently identify "the presence of water" by isolating the local process of interest and performing electrolysis followed by combustion. Our interventions on the candle flame—disturbing the flow of convection currents by blowing on the flame, introducing a spoon above the flame, putting a piece of paper in the center of the flame, etc.—do nothing to disturb the process of electrolysis. Thus, electrolysis is stable under the interventions we perform to learn about the candle flame.[127] Instead of "water"

---

[127] We notice, also, that electrolysis is the process used by Faraday to argue that water is composed of hydrogen and oxygen. For the thing-realist, the composition of water would be the feature of choice for unifying the various dynamics of the candle flame. Thus, by selecting electrolysis as the processual equivalent of the thing-realist's unifying feature, we have effectively translated thing realism directly into process realism with little trouble. I will note here that the process-realist underlier—electrolysis—holds an advantage over the thing realist's underlier—the composition of water. Namely, as we have already discussed, electrolysis is used to learn about the composition of water, not the other way around. The thing realist therefore infers their unifying underlier

we might unify the dynamics of the candle flame using the process of electrolysis (or the ability to perform electrolysis) instead. Similar remarks might be made about the process of interacting with potassium, another of the interventions used to test for the presence of water. In short, the unificatory role of "water" in our explanations can equally be performed by a dynamic feature of the system as it can by a thing-like feature.

The second role that "water" plays is as a labeling tool. I.e., when considering vaporous motion, we might separate this motion into several distinct vaporous motions according to how these component motions differ with respect to our interventions. Notably, one vaporous motion can be affected by a cool spoon or react strongly with potassium while the other remains unaffected by both interventions. We therefore label the affected vaporous motion as the vaporous motion *of water* to linguistically differentiate it from the unaffected vaporous motion (the motion of CO_2 and carbon).

We do this primarily for convenience of reference, not to indicate any ontic feature of the world. Notice, again, that the thing-ness of water does not play an ontic role in isolating this sub-process that we label "the vaporous motion of water." Rather, it plays only a dialectical or linguistic role. It is the intervention of the spoon, and the revealed difference in reaction to this intervention, that separates the vaporous motion of water from the vaporous motion of carbon and CO_2. I.e., we know that vaporous motion of water is an individual dynamic event because this motion can be disturbed by a cool spoon or the interaction with potassium. This motion *is* an individual dynamic event because it leads to different dynamic events in response to interventions than does the motion of the carbon or the CO_2. In short, insofar as "water" plays a differentiating (labeling) role in our explanations, the noun "water" is a placeholder term for the processual differentiators.

The third role that the material presence of water plays is in the connection of non-obviously connected processes. Prima facie, there is no reason to suppose that a combustion process would directly cause a vaporous motion process. After we learn of this, e.g., in a secondary chemistry class, this causal connection seems obvious. However, we learn of this connection by linking the dynamic event of combustion with the dynamic event(s) of the vaporous motion. One of the most powerful ways we can do this is to show that the "end" of one process is identical to the beginning of the other. I.e., we show that the two processes are relatively continu-

---

from dynamic interventions, whereas the process realist simply performs the interventions. In other words, the thing realist requires an additional inferential step from their empirical tests to their ontology (cf. Chapter 2).

ous (cf. Chapter 1), in that they are co-defining such that the ontic and physical features of one are inherited by the other. This leads us to (semi-arbitrarily) separate the single dynamical flow of the system into sub-parts by defining states of the system. We then declare that the state of the system at the end of the first process is identical to the state of the system at the beginning of the next. In our example of the production of water vapor, the stipulated state at the end of the combustion process is defined, in part, by the water that has been produced.

Notice, however, that we are not warranted in claiming that these sub-processes are separate and in need of connection until after we have performed our interventions. The candle as a whole is a single, dynamic event that we observe. This means that their connection is given—it is not something that needs to be explained. Instead, we *describe* the connection of various sub-processes in terms of water and other features of their connecting state(s).

This is a point to which we return in more detail in Chapter 5. For now, suffice it to say that states are defined arbitrarily, by convention, and on top of the actual dynamic events in order to facilitate our understanding of the system. They represent our conceptual apparatus for making sense of an infinitely complex world, not some actual thing or structure of things. Different ways of assigning states take advantage of dynamic, not static, features of the system. Moreover, one state assignment is superior to another in virtue of its usefulness in describing the dynamics of the system and facilitating dynamic explanations of phenomena, not in virtue of any reference to a real thing. This should be, if not obvious, at least plausible given the discussion of this section.

One of the primary objections to a pure process realism is that all systems have material features, and these material features cannot be processes. We now have the tools to understand how this objection fails. Material features arise dynamically, they are explained dynamically, they are learned about dynamically, and, most importantly, they only play an explanatory role in a system insofar as they represent (actual or potential) dynamics themselves. The thing-ness of water is uninteresting. The fact that water can react with potassium is, in contrast, very interesting. Just as with formal features of the candle system, we see that, as far as scientific explanation of phenomena is concerned, dynamics trumps statics.

3.5 Conclusion

We have seen, now, how the features of the candle flame system are all identified with and explained by processes. We also have seen something of a natural pattern emerge amongst those explanations. Productive features are explained by sequential, forked, or jointed dynamic shapes. Structural features are explained by cyclic dynamic shapes. Lastly, material features are explained by no particular dynamic shape, but rather by a comparatively longer time scale or larger characteristic energy than the dynamics of interest in the system. We might therefore say that, for any system, we should expect the following sorts of explanatory processes to appear:

(*Productive or Causal Features of a System*): Those features identified by a generalized continuous sequence of processes, possibly including joints and forks into and out of the sequence.

(*Structural or Formal Features of a System*): Those features identified by a generalized cycle of processes, typically with measurable quantity exchanges that are proportional.

(*Material or Underlier Features of a System*): Those features identified by any generalized intervention process with any dynamic shape, with quantifiable and determinable energetic features (flux, work), that are characteristically larger than or less responsive to the perturbations involved in the system.

Thus, we identify the features of the system as the relevant types of processes we observe and refer to in our explanations. The candle flame is not material in that it has material features, but rather because it contains stable underlying dynamics. The candle flame does not have a structure in virtue of bearing intrinsic relational properties, but rather because it contains cyclic, self-stabilizing systems of dynamics.

It should be noted that this is essentially the task of the process realist: to show that all apparently non-processual aspects of a system or model can be reinterpreted consistently in process-realist terms. The process ontologist must perform this task as well, with a different set of data (linguistic rather than scientific-theoretic). E. g., the GPT seeks to redescribe the truth-makers of thing-, property-, and relational-terms as process truth-makers.[128] If the process realist/ontologist can reproduce all of the relevant scientific explanations and linguistic data that is thought to be captured by thing/substance realism/ontology, then they are no longer subject to the accusation of unmotivated revisionism.

In sum, all features *of the candle flame itself* are successfully explained by, described by, identified with, and defined as processes. At this point, it is possible to resolve the historical tension that motivated this discussion. Namely, there was a tension between:
(1) our current belief that the candle flame is not itself a thing,
(2) that the candle flame (still) possesses certain features,

---

[128] See Seibt 1990, 1996a, b, c, 2004, 2007, 2008, 2010.

(3) that these features of the candle flame were taken as past evidence for reifying the candle flame, resulting in theories of the thing-ness of the candle flame.

The resolution of this tension is simple: the features are still present in the candle system, but they have been redefined by past scientists as a collection of processes (dynamics). I.e., the candle flame is not a thing, but is rather a collection of processes that define its relevant empirical/material features.

We should note an important linguistic point. That is, we still say that the candle flame has these material and formal features. The candle flame is yellow. The candle flame is hot. The candle flame is teardrop-shaped. However, our understanding of these features—the meaning of these propositions—is couched in the underlying dynamics of the system. For example, the candle flame itself does not possess some static property "color." Rather, the system contains dynamics that, in interacting with us and our senses, produce the sensation of color. The color of the candle flame is the interaction itself, not a property in either the candle system or the observer of it (Chirimuuta 2015).[129]

In this discussion, many references were made to the activities *of* other entities, like molecules, fuel, water vapor, etc. On the surface, at least, these terms apparently refer to further things, to "underliers," of the explaining processes. While I have argued that these terms are not explanatory for the features of the candle flame system, it would nevertheless seem that the thing realist has some recourse. They might argue that the candle flame's features are all explained by processes, but that these processes (ontologically) require the existence of further things in order to exist. In other words, while things are explanatorily empty, they are not ontologically empty.[130]

In the next chapter we will consider this argument in more detail. Within the history of physics, the argument itself appears in apparent robustness arguments offered for these new thing terms. In particular, Perrin's argument that molecules must be real within thermal processes is taken in the literature as a robustness argument for things (W. Salmon 1979a, 1981, 2003, 2005). For now, we may note that such robustness arguments are expected to fail. Chapter 2 of this work was precisely designed to articulate how and why no such argument can ever success-

---

[129] This is also a part of W. Seller's work. See Tye (1975) and Vinci (1981) for analyses and defenses of Sellers's adverbialism about sensation. See also DeVries (2005) for a cogent account of W. Seller's work as a whole, including his adverbialism.

[130] One might argue in contrast that things are not explanatorily empty because they are ontologically empty. However, this argument relies on a very strict notion of explanation that only countenances ontological explanations.

fully establish that things, and not processes, are the ontological underliers of processual explanations in science. It should come as no surprise, then, that in the next chapter, I will show just this within the history of physics. I.e., in this chapter, I argued for the explanatory defeat of things, and in the next I will argue their ontological defeat as well.

Chapter 4
Perrin's Argument: A Robustness Argument for Processes, Not Things

4.1 Introduction

In the previous chapter I argued that things do not appear as explanatory features of our models, specifically in the case of the candle flame. This was because all core explanations of the system involve reference to experimental dynamics such as interventions, observations, and in general dynamically contextualized and occurrent features of the system. Our explanations are therefore always made in terms of dynamics, and things need play no role in defining or identifying those dynamics.

Nevertheless, the thing realist persists. Perhaps, they might argue, things cannot be immediately inferred from one particular observation of the candle flame. However, things appear consistently within our explanations and descriptions of particular aspects of the candle flame system and similar systems. Moreover, the same thing terms appear across multifarious descriptions and explanations. Thus, these thing terms (and implicitly, their supposed static referents) are robust features of multiple models. In short, in response to the argument for the explanatory defeat of things in the context of the candle flame, the thing realist offers an underlier argument for the ontological necessity of things. In this chapter, I will show by example how things are defeated ontologically as well.

One of the triumphs of thing realism is the argument for atoms, offered by Perrin in 1909. This argument has been described in W. Salmon (1979c, 1984)[131] as an explicit robustness argument for things. Similarly, Coko (2018, 2019) provides historical support for various portions of Perrin's argument, showing how historical context lends additional strength to the so-called robustness argument for atoms. Many other accounts treat the case similarly (Brush 1968; Chalmers 2009, 2011; Nye 1972, 1984).[132] While each of these accounts is subtly different in the details, the basics of the argument are the same. Namely, the thing realist argues that in Perrin's work (and also Einstein's preceding work on Brownian motion), atoms

---

[131] See also W. Salmon's other work for context and for further references to this robustness analysis. E. g., W. Salmon (1990, 1994, 2005). The connection with Reichenbach is informative and can be found in W. Salmon (1979a, b, 1994).

[132] For similar discussions of Perrin's arguments that are less obviously thing-realist, see Cartwright (1991), Glymour (1975), Mayo (1986, 1996), and Psillos (2011a, b).

appear as robust posits in many different models and calculations of thermodynamic systems. They are therefore real entities.

It does not take much to see that the thing realist has a good argument here. Not only are atoms apparently robust across models, they apparently have the properties of thing-like entities. Namely, they have definite, atemporal properties, are particular in the sense that they are spatiotemporally definitely located, are countable, and are independent of dynamic or experimental context. This last point, especially, follows from their being robust. At least, so goes the argument.

However, as I argue in this chapter, this supposed triumph of thing realism is mistaken. Perrin's arguments can and should be understood as arguments for specific dynamic entities, not for things at all. The argument for this processual conclusion requires a close reading of Perrin's 1909 and 1916 arguments in particular. In these, we will see that Perrin is offering a means of consistently identifying the measurable aspects of thermal dispersion processes. Importantly, these dispersion processes come in specific countable amounts with a characteristic (but only determinable!) size, but are not themselves particulars. Therefore, we see in Perrin's work an argument that thermal dispersion processes have a characteristic measurable size, not an argument for atoms qua things.

We proceed as follows. We begin with Perrin's intuition pumping example from his 1916 work (§4.2). This sets the stage for the explicit qualitative arguments and calculational arguments to come. After considering these more detailed arguments in (§4.3) and (§4.4) respectively, we conclude by considering some general points from later atomic physics, specifically quantum mechanics, that show how physicists following in the footsteps of Perrin were explicit in their moves toward interpreting physical systems in terms of non-particular, dynamically contextual, determinables rather than determinate particular independent things (§4.5). This will close our chapter on the ontological defeat of things.

4.2 Perrin's Intuition Pump: The Bath and Cascading Fluid Motion

Perrin is often cited as the, or one of the scientists who proved the existence of molecules through the analysis of Brownian motion.[133] His arguments in 1909, and later 1916, are taken to be paradigm examples of so-called robustness reasoning. The intuition is that, by showing how to calculate Avogadro's number through multiple experimental and mathematical methods, Perrin established that this

---

[133] Another oft-cited scientist in this regard is Einstein. See Einstein (1905a, b, c, d, e).

number was a fundamental parameter in the world, and thus was an actual count of something. The close agreement of the values calculated, and the independence of the calculation/discovery methods, acts as evidence that this number is no coincidence or artifact of our experience. Rather, it is a *statistically relevant* aspect of the world, to use Salmon's 1970 phrase (cf. Reichenbach 1956). This is taken to mean that the things being counted are statistically relevant, and therefore explanatory, elements of reality. Implicitly, the standard interpretation of this argument requires that only things may be counted.[134]

While some have questioned the exact nature of Perrin's argument (cf. Hudson 2014, 2020) who argues that Perrin is a "calibration" reasoner, not a robustness reasoner), Perrin's argument is always taken as an argument for things, molecules/atoms in particular. However, analysis of Perrin's work shows this to be mistaken. Perrin begins his 1916 work with an intuition pump for the importance of Brownian motion:

> When we consider a fluid mass in equilibrium, for example some water in a glass, all the parts of the mass appear completely motionless to us. If we put into it an object of greater density it falls.... When at the bottom, as is well known, it does not tend again to rise, and this is one way of enunciating Carnot's principle.... These familiar ideas, however, only hold good for the scale of size to which our organism is accustomed, and the simple use of the microscope *suffices to impress on us new ones which substitute a kinetic for the old static conception of the fluid state.* (Perrin 1916, 1, italics mine)

Perrin's intuition pump goes as follows: we see in a fluid a particular stability, a lack of motion. However, investigating this reveals not that there is no motion, but rather that the motion is merely balanced, evenly distributed throughout the fluid, such that it escapes our ability to immediately observe it.

In other words, Perrin's opening framing device sets the stage for his investigation into the underlying kinetics (read: dynamics) of the fluid, not into its thing-like composition. Already, this indicates that Perrin's argument for the existence of molecules is not quite the paradigm of thing realism that Salmon's gloss suggests, since the goal of Perrin's analysis is to arrive at a processual understanding of the fluid. The question, then, is whether this processual understanding of the fluid requires the existence of things.

Perrin's argument for molecules/atoms then takes three forms, one qualitative followed by two calculational arguments. The latter arguments are the true argu-

---

[134] As we have seen in Chapter 1, processes are indeed not countable, but can come in amounts that are quantifiable. These quantified amounts (1.000 lumens in a radiative process, 200 joules in this motion, etc.) can then be counted.

ments for the existence of molecules, and occupies the remainder of his paper. The former argument within pages 2–7 is an eliminative argument that the cause of Brownian motion is the motions of the parts of the liquid transferring their motion to the suspended particle. It is notable that this argument appears first, before the argument for the existence of molecules. The eliminative argument must succeed in order to provide (a portion of) the necessary evidentiary base for the second argument, as we will see. We turn now to this first argument, the qualitative one.

4.3 Perrin's Historical/Eliminative Argument

Perrin's opening eliminative argument begins on page 2. In the eliminative argument, Perrin cites the major experiments and arguments of his predecessors that rule out other possible causes of the motion of the suspended particle. First among these is Weiner, whom Perrin cites (1916, 2) as arguing first that the motion of the suspended particle cannot be due to currents in the air or to currents in the fluid arising from thermal disequilibrium (the standard explanations of the time). Father's Delsaulx and Carbonelle, Perrin quotes as arguing first that Brownian motion is indicative of some general property of the matter composing the fluid in which is suspended the Brownian particle, namely that the fluid is composed of corpuscles. However, Perrin suggests that these were all ignored because their work was "superficial" (1916, 3), lacking proper experimental test.

It wasn't until Louis Georges Gouy in 1888 that the standard explanation of Brownian motion was truly questioned, according to Perrin (1916, 4–5).[135] In his work, Gouy provided a series of experiments to test and rule out all but one explanation of Brownian motion in a paradigm example of eliminative reasoning. First, Gouy rules out that the motion of the suspended particle is caused by convection currents in the fluid by testing to see if the motion changes after enough time has passed for the fluid to reach thermal equilibrium. Finding that there is no difference in the motion of the particle before and after this equilibration period, Gouy rules out that the thermal state of the fluid accounts for the motion.[136] Next, Gouy rules out that Brownian motion is caused by the external transmission of motion through, e. g., vibration impacting the fluid container. He took two equally prepared systems and observed one at night in the countryside and the other during the day

---

[135] It is probably significant for the dissemination of Gouy's arguments that Maede Bache provided confirmation for Gouy in 1894.
[136] We also know, according to Gouy, that this is a different kind of motion than, e. g., the motion of dust particles in air because the movements of two Brownian bodies are completely independent of each other, whereas the dust particles exhibit a common coherence in their motion.

in a busy London street. Finding, again, that there was no difference in the motion of the suspended particle, Gouy ruled out that external vibrations could be causing the motion. Next, Gouy rules out that the cause of Brownian motion is light from any source impinging on the particle, what Perrin calls "unavoidable illumination" (1916, 5). To test this, Gouy performed a series of rapid and radical changes to the color and intensity of the light impinging on the suspended particle. Again, this produced no difference in the Brownian motion. Lastly, Gouy makes use of Brown's own arguments to show that the nature of the suspended particle has no bearing on the motion.[137]

Perrin takes this eliminative argument of Gouy's as definitive evidence that the cause of Brownian motion lies in some fundamental feature of the fluid:

> Thus comes into evidence, in what is termed a *fluid in equilibrium*, a property eternal and profound. This equilibrium exists as an average and for large masses; it is a statistical equilibrium. In reality the whole fluid is agitated indefinitely and *spontaneously* by motions the more violent and rapid the smaller the portion taken into account; the statical notion of equilibrium is completely illusory. (Perrin 1916, 5–6, Perrin's emphasis)

According to Perrin, even when in thermal equilibrium, with no external sources of motion, the fluid exhibits a kinetic nature that is fundamental to all matter. It is this kinetic nature that is the cause of Brownian motion. Equivalently, the suspension of a particle within a fluid acts as a probe of this fundamental kinetic nature of the fluid.

This, then, is the eliminative argument that Perrin provides. We begin by considering an observed phenomenon: the motion of the suspended particle. We notice that this motion involves many small and sudden changes in direction and speed in the suspended particle. We then ask: what is the cause of these changes? We then construct a list of possible causes of the changes in motion. We know from previous investigations that changes in momentum require a transfer of momentum from some other source.[138] Thus, there must be motion external to the suspended particle that can transfer motion to the particle (the motion of the particle is not spontaneous). Upon looking at the viable sources of motion—external vibration, illumination, thermal disequilibrium, air currents, the internal character of the suspended particle—we discover that none of these makes any sort of difference to the motion of the suspended particle. Only the nature of the fluid in which

---

137 See Brown (1828) for details. Brown's experimental procedure is another example of eliminative reasoning.
138 Note that this is an assumption about macroscopic and minimally microscopic systems and phenomena. It is well established, but it is an assumption all the same. Phenomena on the quantum scale require a slight amendment of this assumption, as we will see in later sections.

the particle is suspended makes any difference. We therefore conclude that the motion of the particle must come from some intrinsic motion of the fluid that is not an aggregate motion like, e.g., the motion arising from thermal disequilibrium.

We investigate each of these sources of motion in much the same way we investigated the features of the candle flame, i.e., through a series of interventions meant to isolate the motion (and source of motion) of interest. E.g., we intervene on the fluid + particle system to isolate external vibrations as a potential source of motion by placing identical systems in different locations, one with a much external vibration and the other with little. The only difference between these being the impact of external vibrations, any difference we observe (any change in the character of the Brownian motion) must be due to the external vibrations. Observing no such change, we conclude that external vibrations are not a cause of Brownian motion. Similarly, we intervene on the fluid+particle system to isolate the nature of the fluid as a potential cause by altering the fluid in which is suspended the Brownian particle, e.g., by placing the same particle in water and then in a more viscous fluid. In contrast to the case of external vibrations, we observe a difference in the Brownian motion of the particle in this case. Since the only difference between these two systems was the characteristic dispersion, or non-aggregate motion, of the fluid, we infer that this characteristic dispersion of the fluid is a cause of the Brownian motion we observe.[139] By eliminating all but one such source of motion through these interventions, we infer that the internal motion of the fluid is the sole cause of Brownian motion.

This eliminative argument, however, is not truly an argument for the reality of atoms. We have not discovered, through this analysis, any feature of the fluid that could only exist given the existence of a thing-like entity to compose the fluid. Rather, we have established only that the cause of the motion we perceive in the suspended particle is the aggregate transfer of motion from the components of water, and that these motive components are many times smaller than the suspended particle that is caused to move.[140] Equivalently, we have shown that the fluid exhibits a characteristic dispersion of motion, and this dispersion is the

---

[139] Note, anachronistically, that we can perform additional interventions on the fluid to test *which* internal aspect of the fluid is a cause. E.g., we can compare viscous to non-viscous fluid, polar to non-polar, and Newtonian vs. non-Newtonian.

[140] Note, here, that there is a difference between saying "the aggregate transfer of motion" and "the transfer of aggregate motion." The latter would refer to the transfer of motion from, e.g., a current in the fluid, which is an aggregate motion. The former refers to many individual transfers of motion.

cause of the motion of the suspended particle.[141] It is only when we further infer that this characteristic dispersion is indicative of the sum of motions of constituent things—molecules—that we are able to infer the existence of those things. This is the purpose of Perrin's second argument: to show that the characteristic dispersion is indeed the type of dispersion one gets in a substance that is composed of an aggregate of molecules.

We will turn to this second argument in a moment. Before we do, notice that we cannot provide an elimination argument to the effect that the characteristic dispersion of the fluid is a dispersion of the motions of things. I.e., we cannot list every dispersion rate equation, label each as a dispersion rate of things or a dispersion rate of processes, and then show that the dispersion rate in the fluid is a dispersion rate of things not processes. This is because the difference between the process and the thing ontology lies not in the particular terms of the dispersion rate equation but in the emphasis placed on the equation as a whole. This suggests that the priority of things over process must be put into our arguments by hand, rather than arising from the mathematics alone. In contrast, the eliminative argument we have just discussed establishes that Brownian motion—a process itself—is caused (brought about, produced) by a characteristic process of the fluid in which is suspended the Brownian particle: the dispersion of motion. In short, just like we found in the candle flame, our inferences to the existence of processes arises as a direct result of our interventions and observations, whereas our inferences to things require additional assumptions. We now discuss Perrin's second argument, where this will become even more apparent.

4.4 Perrin's Precise Arguments

4.4.1 Argument 1: Qualitative Robustness

Having shown that there is some characteristic dispersion of the fluid, Perrin moves to provide an argument that Brownian motion proves (or requires) the existence of molecules. That is, Perrin provides various derivations of Avogadro's number, N. The sum of these derivations, their diversity and independence, is

---

[141] It may not be obvious why the claim that there is non-aggregate motion internal to the fluid is equivalent to the claim that the fluid has a characteristic dispersion. The connection lies in the fact that non-aggregate motion is the cause of equilibration in a fluid. This means that, whenever the fluid is in disequilibrium with itself, in which case it will exhibit aggregate motion of some kind, it will equilibrate through the loss of aggregate motion by way of non-aggregate motion. The loss of aggregate motion is identical (in this case) to diffusion.

then meant to provide reason to believe that N is a fundamental feature of the fluid in which the Brownian particle is suspended. Since in all of these derivations N arises as a consequence of observing the interaction of the fluid and the Brownian particle, N must represent some feature of the cause of Brownian motion. Since N is a whole number related to weight and density, N is taken to count things—molecules—whose collisions with the Brownian particle cause it to move in the characteristic erratic way. Thus, Brownian motion must be caused by the existence of molecules that move independently of each other. Perrin then adds the further argument that, since the value of N can be calculated in a multitude of ways, our apparent ability to count things within the fluid is no fluke. I.e., N is a fundamental feature of reality because it is robust to differences in experimental and calculational methods, including different idealizations and approximations. This is what later interlocutors term Perrin's "robustness" argument.

Interestingly, this robustness argument opens first with a qualitative argument that the fluid is composed of *some* entity. The eliminative argument establishes that the cause of Brownian motion is some feature of the fluid itself. However, this alone is not reason enough to suspect that this feature of the fluid is a feature of some *fundamental constituent* of the fluid. Perrin does, eventually, infer that the cause of Brownian motion is the fundamental motions of the constituents of the fluid (and their dispersion through collisions). But notice that this inference requires an additional premise: that the fluid may be decomposed into more fundamental constituents. Whether these constituents are constituent *motions* (processes) or constituent *molecules* (things) remains undecided, but Perrin must suggest to the reader that the fluid has constituents. Otherwise, Brownian motion, which our elimination argument established is caused by some feature of the fluid, could be caused by some intrinsic feature in the fluid, thus indicating no further decomposition of the fluid into individual constituents.

For example, Brownian motion could be caused by spontaneous motion in the fluid. Such motion would not be aggregate motion of the fluid, which Gouy ruled out when he showed that thermal currents in the fluid are not the cause of Brownian motion. However, spontaneous motion in the fluid could still be a feature of the fluid as a whole and also the cause of motion in the Brownian particle. In such a case, we would say that the cause of Brownian motion is the tendency of the fluid to spontaneously acquire localized momentum (with some decrease in momentum elsewhere to compensate). Importantly, the location of this acquired moment, the region in which it arises, would have to have no definite length scale.

Perrin therefore needs to establish that there is some characteristic constituent of the fluid *before he can calculate N in the first place.* In other words, Perrin must show that there is a regularity in the fluid suggestive of the existence of some entity out of which the fluid as a whole is composed. He does this by establishing that the dispersion of the fluid which is the cause of Brownian motion has a characteristic, observable size. Put simply, dispersion of motion in the fluid does not occur indefinitely, but only until motions within the fluid occur on a particular length scale. Once the motions are all on that length scale, dispersion no longer produces de-coordination in the motion of the fluid.

This notion of de-coordination is key. To illustrate it, Perrin considers a large amount of water poured into a bathtub (1916, 8). The water, Perrin remarks, will exhibit coordinated motion: any two regions of the fluid will be moving in approximately the same direction with the same speed. This coordinated motion corresponds to what we might anachronistically call aggregate motion in the water, and can be observed by introducing colored powder into the water and tracking the trajectories of the colored regions. After the initial flow of water into the tub, the water will collide with the walls of the tub. The water will then exhibit motions that are still coordinated, but on a smaller length scale. Roughly, a portion of the water will splash left, another will splash right. All regions of the left-splashing water will remain coordinated in their motion, but the left- and right-splashing water will no longer be coordinated. This is the de-coordination that Perrin is interested in: the coordination between motions in the water exists *within smaller regions of the water.*

Perrin then asks: is this de-coordination of the water never ending? I.e., do the coordination regions continue to shrink in size indefinitely? What we find is that they do not shrink indefinitely. Rather, de-coordination occurs only until the motions within the water occur at a particular length scale. At this point, every further de-coordination of the motion within two regions of the water will incur an equal and opposite coordination of each of these regions with another region. I.e., if X and Y de-coordinate, X and re-coordinates with Z_1 and Y with Z_2. This is equivalent in modern parlance with saying that dispersion of the water's non-thermal energy brings the water into equilibrium. Again, we can observe this (according to Perrin) by introducing colored powder into the water and observing the motion of the colored region.

Since the coordination in the water's motion will always exist at a certain length scale, Perrin infers that the water is composed of *some* entity of that size.[142] The natural entity to posit is a thing—molecules—since N is a counting number. However, it should be noted at this point that Perrin makes no pronounce-

---

[142] Note that this is a move from dispersion to scattering in Perrin's thinking, since he immediately invokes thing-like constituents of the fluid and their collisions to explain these de-coordinations.

ment that would indicate that N is any more special than other fundamental parameters that one can calculate for fluid (and gas) systems. Importantly, this includes k, Boltzmann's constant.

Having established that the dispersion of the water exhibits a characteristic length scale, and that this length scale suggests that the constituents of the water are of that length scale, Perrin is ready to argue that N is a fundamental parameter counting molecules in the world. We will discuss his first calculation in the next section. Before we do, we must note that Perrin's argument manifestly rests on the observation of dispersion (de-coordination, in his language). As far as his experiments are concerned, dispersion is the phenomenon he can actually manipulate, intervene upon, and observe. Dispersion is a process, as are the microscopic motions that Perrin appeals to in his account of the de-coordination of aggregate motion in the water. Therefore, Perrin's argument follows the same pattern of inference we have been noting all throughout this chapter: we begin with observed dynamics, infer further dynamics through the use of careful intervention and observation, and only then infer that there exists some underlying thing for these dynamics. Thus, while Perrin argues that molecules can be inferred from these observations, we must always remember that this is an inference several steps removed from the bare observation of the system.

This is something of which Perrin seems aware. While he argues definitively that N is a fundamental parameter in the world, and thus believes that there is something being counted, he is reticent to draw any further conclusions from this fact. In particular, even though the existence of molecules would naturally suggest that collisions between them are the cause of diffusion of their motion, Perrin remarks that redistribution of motion within the fluid occurs "by *impact* or in any other manner" (1916, 9). This sort of reticence indicates that Perrin is aware that, while N can be said to refer, the metaphysical nature of its referent is not thereby established. This proves an invaluable point to exploit for the process realist.

4.4.2 Argument 2: Quantitative Robustness

Here, we consider Perrin's calculations of N. In these calculations, we note the assumptions that Perrin makes in order to perform the calculations. Namely, Perrin assumes something about the size of the molecules in the fluid and something about their density. Finding that these assumptions are warranted on the basis of a dynamic analysis of the system, or else are assumptions about these dynamics, I argue that Perrin's calculations can be restructured into an argument for processual underliers of diffusion rather than thing underliers. This restructuring is quite simple, and makes obvious the ways in which the calculations represent inferences from experimental tests. We conclude that Perrin's argument is not an argument for things at all. Indeed, it seems more natural to interpret it as an argument for a processual decomposition of an observed macroscopic, dynamic phenomenon. In short, Perrin's argument becomes an argument for the existence of things only if we accept Perrin's interpreters' thing-ontological framing of the calculations in the first place. If we simply reframe the calculations, they become arguments for processes.

4.4.2.1 The First Calculation: An Estimation

Perrin's first calculation of N (1916, 13–18) is an attempt to estimate N by establishing an upper and lower bound on N's value. To begin, Perrin notes that previous work by Maxwell and Clausius establishes a relationship between the density of molecules in a given volume and their average size.[143] Namely, assuming that molecules in a gas are approximately spherical, we have

$$L = \frac{1}{\pi\sqrt{2}} \frac{1}{nD^2}$$

where L is the mean free path that a molecule traverses between any two collisions within a gas, n is the molecular density equal to N divided by the volume V, and D is the diameter of a molecule approximately the shape of a sphere. Colloquially, this means that spherically shaped molecules will collide with each other every L units of distance they travel, and that this distance decreases as the molecules increase in size and packing within a given volume.

L is calculated independently. We assume that the molecules in a gas have speeds in the x-coordinate direction distributed according to the Gaussian:[144]

$$\frac{1}{U}\sqrt{\frac{3}{2\pi}} e^{-\frac{3x^2}{2U^2}} dx$$

Here, U is the root mean square (rms) velocity of the molecules which is determined solely by the temperature of the gas. (Note that the variable x is referring to speeds, not coordinate positions.) If this is the case, it follows that the average speed of a molecule, Ω, in the gas is:

$$\Omega = U\sqrt{\frac{8}{3\pi}}$$

---

[143] Perrin is referring to Maxwell (1860) and Clausius (1851).
[144] This Gaussian distribution will hold if the molecules in the gas are assumed to be roughly non-interacting, independent constituents.

Combining this with a measurement of the viscosity ξ (i.e., the internal friction of the gas) and the absolute density of the gas, δ, we can determine that:

$$\xi = 0.31\delta\Omega L$$

δ is measured using a finely tuned balance. ξ is calculated as a solution to a differential equation for the angular velocities in a specially prepared system of coaxial cylinders containing the gas. Put simply, we determine viscosity by noting the relationship between the changes in angular speed of the inner cylinder when the outer is driven to rotate by a motor.[145] Since the gas is the medium through which the angular speed of the outer cylinder is translated, the inner cylinder will acquire angular speed at a rate proportional to the ability of the gas to translate motion, and inversely proportional to the gas's resistance to motion.

What remains is to establish an additional relationship between n and D so that n, and thereby N, can be calculated. Perrin provides two such relations, namely two inequalities, and thereby provides an upper and lower bound to N. The first relation assumes that the gas molecules can be no more closely packed than billiard balls in a closed container (1916, 15), thereby providing an upper bound on the density of the molecules. The second relation assumes that the molecules can be no less closely packed than a number of perfectly conducting spheres would be in a volume with the same dielectric constant as the gas (1916, 16), thereby providing a lower bound on the density on the molecules. Roughly, this allows Perrin to calculate that N must be between 4.5 and 20 times 10^{23} (the true value is 6.022 times 10^{23}).

Notice that this estimation of N cannot be taken as any evidence for the existence of molecules as it has been written by Perrin. The language used so far, e.g., that the gas contains molecules of diameter D, has assumed that molecules exist a priori. If we were to argue that there are N things (molecules) contained in a gas because a gas consisting of molecules with diameter D will contain N molecules in a volume V, we would clearly be begging the question. Instead, if we wish this to act

---

145 One possible system works as follows: we contain a gas within the space between two coaxial cylinders. The inner is allowed to rotate freely on a torsion fiber with a fixed support. The outer is rotated at a constant speed w by a motor. This introduces a velocity gradient within the gas, w at the outer boundary and 0 at the inner. The calculation then proceeds by measuring how quickly the inner cylinder acquires angular speed.

as an argument that the gas really does consist of things, and that there are N of them in a mole, we must reinterpret the language used in Perrin's estimation.[146]

This reinterpretation is rather simple. All we must do is replace phrases such as "D, the diameter of the molecule," "L, the mean free path between collisions of two molecules," and "Ω, the average velocity of a molecule in the gas" with their empirical grounds. For example, we calculate Ω by noting that, at a given temperature T, an enclosed gas imparts a certain amount of energy to the surface of its container. This allows us to calculate U, and thereby calculate the distribution of velocities. From this distribution, we calculate Ω, the average speed. In this way, we link Ω with the measurable values of the temperature of the gas and the average energy imparted to the container of the gas per unit time.

Importantly, within these calculations, a certain granularity appears. This is the same granularity to which Perrin appeals in the qualitative argument discussed above. Namely, the redistribution of motion/energy within the gas system exhibits a characteristic length scale when it reaches equilibrium. In Perrin's analysis, this granularity is expressed in the number of molecules, the amount of substance, in the gas. However, notice that we could equally remark that there is a granularity in the energy transferred to the container of the gas per unit time, and therefore granularity in the energy of the motions observed in the gas. Namely, the mean energy of each motion turns out to be:

$$E = \frac{3}{2}KT$$

This is typically interpreted as meaning that each molecule in a (monatomic) gas possesses approximately 3/2KT units of energy, plus or minus some degree of variance from the mean. This variance is then taken to be the cause of the random distribution of motion in the gas, and so the cause of the random motions of a Brownian particle. However, if we are arguing that N represents a number of things in the world, we are not permitted to perform this interpretation. Instead, we must simply say that there is a granularity in the energy of motions in the gas.

This reinterpretation must be performed for every calculation that follows in Perrin's analysis. Crucially, in performing this reinterpretation in Perrin's later calculations, we will discover that for everything Perrin says about N, we can find an equivalent statement about K, the granularity in the energy of diffusion. K is, prima facie, a property of processes (it is a measure of the size of a "collision" in-

---

[146] Consider also: the natural way to determine n is to count the number of molecules in a given volume. However, we cannot do this practically. Instead, we must use dispersion and scattering analysis in order to calculate n.

teraction). Thus, we have parity between a process interpretation of Perrin's calculations and a thing interpretation. This parity, resulting from the calculational parity between K and N, is something Perrin is aware of (1916, 11, 18). We turn now to Perrin's second, more precise calculation of N, in which these points become even more obvious.

4.4.2.2 The Second Calculation: More Precisely

To achieve a more exact calculation of N, Perrin needs only to make the method of calculation more precise. His goal is to provide a sequence of calculations meant to establish the connection between the relevant constants N and k, and experimental parameters that can be accurately determined. By doing this, he can specify multiple experimental means of determining these experimental parameters, thereby determining "in the same step the three universal constants N, e, and α ..." (1916, 12) in virtue of their relationship with these experimental parameters. (Note that α is a stand-in for k in this instance.) His remaining calculations of N, therefore, rest on merely describing different experimental means of filling in the details of this relationship. Crucially, the majority of Perrin's effort is therefore placed not in the mathematical calculations, but in the experimental justification for the application of these mathematical relations. In parallel with this, calculations in this section will be omitted in favor of discussion of the more relevant experimental justifications.

Perrin begins by calculating the mean square speed of the molecules in the gas. This is done by simply noting that, for ideal gases, the mean square speed is equal to the square root of R (a known constant) multiplied by the temperature of the gas (which is determinable by direct intervention on the gas). We then calculate, following Maxwell, the distribution of speeds around this mean. Assuming that the component speeds are independent, we find that the molecular speeds vary in a Gaussian:

$$\frac{1}{U}\sqrt{\frac{3}{2\pi}}e^{\frac{3x^2}{2U^2}}dx \frac{1}{U}\sqrt{\frac{3}{2\pi}}e^{\frac{3x^2}{2U^2}}dx$$

From this, we obtain the mean speed of the molecule, which differs from U by a multiplicative constant.[147]

Perrin then needs to calculate the mean free path of the molecule, or the mean distance traveled between collisions. This will depend on the shapes of the collid-

---

[147] Maxwell uses this to determine the viscosity of a gas by showing that the friction between two planes of the gas is the result of constant exchange of speed/kinetic energy. The result that this viscosity is unrelated to the density of the gas is, according to Maxwell, quite surprising.

ing molecules, but can be approximated experimentally by measuring the average distance between two colliding molecules upon collision, the impact parameter I, and treating I as the radius of the molecule. This distance is measured by observing the angular change in the trajectory of a molecule in a collision event, which is done within a bubble chamber. Having obtained I, we then calculate the mean free path of the molecules (following the work of Clausius 1851) as being inversely proportional to I and n, the density of molecules. Perrin notes that a second relation between I and n must be given to facilitate this calculation of L, since neither I nor n is independently known. By obtaining a second relation between I and n, we can calculate n, and then multiply by the known molecular weight to obtain N (thereby obtaining the three constants N, e, α). Perrin proceeds (pages 15 and on) to provide various ways of obtaining this second relationship.

Once again, we must stress that Perrin's argument cannot yet be taken as an argument for things. First, we must substitute experimental values for terms such as "the radius of the molecule" and "the density of molecules." Without this substitution, Perrin's argument would be viciously circular, seeking to prove that there are such molecules while assuming brute facts about their existence. Thus, we must once more read his remarks about the size of the molecule and the molecular density as statements about the experimental method of determining these and related parameters.

This means we are now in a position to understand the true workings of Perrin's analysis. Perrin is explicit in remarking that the impact parameter is obtained through the analysis of experimental processes, i.e., through the angular deflection of observed motions in a bubble chamber. The question, then, is how we are meant to interpret the molecular density n in terms of experimental observations. Luckily, this, too, is not terribly difficult, but requires us to depart from Perrin's explicit analysis slightly. In particular, we must replace n with a statement about the characteristic time between collisions in the fluid, i.e., the dispersion rate. It is for this reason that Perrin's opening, qualitative description of the fluid system is so crucial to his analysis: it provides the necessary intuitions relating the dispersion of the fluid's aggregate motion and the density of the component motions.

To calculate this characteristic time is similarly simple, following the experimental set-up of the qualitative argument with which Perrin begins the paper. As an intuition pump, if we introduce into the liquid a particle with a known mesoscopic diameter—a Brownian particle—we can measure the average impulse provided to it by measuring its changes in momentum across some set period of time. I.e., each change in momentum will be supplied by some number of collisions, or rather, some accumulation of kinetic energy per unit time. This corresponds to some loss in kinetic energy per unit time—dispersion—within the fluid, as the fluid transmits its kinetic energy to the Brownian particle. This would allow us

4.4 Perrin's Precise Arguments — 141

to determine through measurements of times/durations alone a simple relation between the size of the Brownian particle and the dispersion, from which we could estimate the number of collisions per unit time. With our previous assumptions about I, the impact parameter of the molecules striking the Brownian particle, and a few additional reasonable assumptions about the distribution of momenta for those molecules, we could then calculate n, the molecular density of the fluid. Note this is essentially what Perrin does in section 14 (page 23 on). He then describes how one actually sets up this experiment in later sections (e.g., section 16 on page 27).

In other words, the molecular density, n, is calculated in terms of measurable times for dispersion processes, assuming a particular distribution of motion (i.e., the Maxwell distribution) in the fluid and a relation between the relative impact parameters of the relevant interacting bodies. From this, we calculate the constants, N, e, and α, with α being the first and easiest to calculate. The importance of this cannot be overstated. In providing a calculation of n, and thereby of N, Perrin is forced to first determine experimentally more fundamental parameters of the system, namely, average dispersion of the fluid and the average interaction strength and angular deflection in each dispersive collision (i.e., the impact parameter). In other words, we have replaced thing-like terms with parameters and terms that refer to dynamics, properties of dynamics, and interactions thereof. Put another way, Perrin's calculation of N can be co-opted and transformed into a calculation of k, the average dispersion of motion in a thermal system.

As Perrin remarks (1916, 18) there is equivalence between the claim that the partial pressure of each gas in a mixture remains fixed and the claim that the mean kinetic energy of each gas in a mixture remains fixed. The former is treated as a feature of the density of the gas, but the latter is clearly a processual property. I.e., the measure of the fundamental unit of motion within each gas or fluid remains constant. Similarly, Perrin notes that N (the count of component particulars) and k (the measure of component motions) are equivalent from a calculational point of view (1916, 11). The question is not which is fundamental. Rather we are driven to choose one over the other in virtue of either our pre-existing ontological commitments or else in virtue of the priority of one over the other in experimental measurement practices.

Perrin tips the scales in favor of the thing realist by beginning his analysis with a discussion of N as a fundamental constant. However, the fact remains that the mean kinetic energy of translation of different molecules can be used to calculate N. In fact, the result of our calculations is that, rather than revealing some fundamental *thing-like* property of fluids through analysis of Brownian motion, we have revealed only that there is some countable feature of the fluid. In process-realist terms, we would say that every fluid has a fixed parameter defining the rate at which dispersion occurs, and that this dispersion translates into a countable accumulation of kinetic energy of a Brownian particle suspended in the fluid. Moreover, this dispersion relation is the key qualitative indication of this countable feature of the fluid, as evidenced by Perrin's opening qualitative description. It is this priority in experiment that breaks the calculational parity between N and k.

4.4.2.3 Breaking Parity between N and K

Manifestly, the process is prior to the thing in experiment. We do not measure Avogadro's number. We calculate it from measurements of dispersion, i.e., from measurements of observed processes and their interactions. In other words, the thing is inferred, but on what grounds? Clearly, it cannot be on experimental grounds alone, otherwise we would not have an equivalent process-realist interpretation of Perrin's counting argument in the first place. Rather, we require additional assumptions to make Perrin's argument into an argument for real things. The character of these assumptions is the subject of this subsection. What we will find is that these assumptions come from little more than a desire for the world to be composed of things and substances in the first place. This presupposition can only ever beg the question against the process realist, and so the thing realist is forced to turn to other arguments to solidify their ontology.

The first means of transforming Perrin's argument into an argument for the reality of things is to assume that only things can be counted. If this were true, then the existence of a term in the equations of our models that takes only whole number values indicates the existence of things to be counted. An argument similar to this is used in the context of quantum mechanics. Namely, some interlocutors argue that the existence of a number operator in certain quantum systems is an indicator of haeccity in the components of the quantum system. Moreover, it is argued that, since the number operator is not preserved in certain systems (e.g., in a system consisting of electromagnetic radiation in a mirrored container), a clear distinction can be made on this basis between quantum systems containing things and quantum systems that do not contain things.

However, the unstated premise of these arguments—that only things may be counted or come in whole numbered amounts—is not sufficient. Process types *can be counted*, and process tokens can be measured and thereby quantified. If I see two candles before me, I say that there are two (type) processes of combustion before me. If I see two balls rolling on a table, I say there are two (type) motions across the table. If I consider this dispersion processes, I can measure it in units of $3/2kT$, and thereby quantifiably scale it against other dispersion processes, or divide it into intervals of dispersion characterized by a particular multiple of $3/2kT$.

Moreover, in Perrin's argument, the process-realist redescription of a thermal system exhibits granularity just like the thing interpretation. The difference resides in where and how this granularity appears in each respective interpretation. In the thing interpretation, the granularity appears as a result of the nature of the material constituents of the fluid, namely the molecules. In the process interpretation, the granularity appears as a result of the characteristic size of the typical interactions within the fluid. Namely, all motions within the fluid interact with an energy approximately equal to a whole number multiple of the mean energy $3/2kT$.

Admittedly, processes are uncountable. We therefore only ever count numbers of certain semi-arbitrarily determined amounts of quantities carried by processes. Processes exhibit a continuous character that does not lend itself to non-arbitrary counting of them. In part, this is because processes are by nature not fully separable from their environment. One may therefore attempt to justify the premise that only things may be counted by claiming that only things can be discrete individuals. I.e., only things have a character that permits them being identified in the world as wholly separate from their environment. If this were the case, then only things could be counted by whole numbers *non-arbitrarily*. This then is our second means of transforming Perrin's argument into an explicit argument for things.

Prima facie, this revised transformation of Perrin's argument stands on firmer ground. However, even modified, the premise of this argument is problematic. Firstly, things are not wholly separate from their environment, as all things must be defined partly by their relation to their environment. In particular, the identification of a discrete individual thing no doubt relies on the location of that individual. Absolute spatiotemporal coordinates being impermissible, the location of a thing must be defined relationally.

More importantly, insofar as things are separable from their environment, so too are processes. Without desiring to be too technical, things are individuated and separated from their environment by the recognition of differences between the thing and the environment. The water glass is different, and thus a separate individual, from the table it sits on because it is a different color, has different reflective properties, etc. The same can be said of processes. Combustion is different from radiation in the candle system because radiation interacts with the wax of the candle to melt it whereas combustion does not, radiation triggers color receptors in our eye and combustion does not, etc.

In fairness, I am not attempting to argue that processes are discrete individuals in the same sense as are things. It seems to me that one of the primary differences between a process and a thing ontology is exactly this point: that processes are not discrete individuals fully separable from their environment as are things. However, it is important to note that this difference is a metaphysical difference between the two ontic units. The discreteness of a thing comes not from our means of identifying it empirically, nor our means of defining it, but from some a priori stipulation that the essence of a thing is non-relational. This means that, to appeal to this discreteness in Perrin's argument for the sake of establishing the existence of things is to appeal to a metaphysical principle without empirical ground. Once again, we find that the argument Perrin provides for molecules exhibits parity between a process and thing interpretation.[148] This parity is broken by metaphysical, not empirical, premises.

The third and final means of transforming Perrin's argument is to declare that the dynamics of diffusion requires that there are at least two things to deflect each other's motion. I.e., the processes involved in Brownian motion require thing underliers. This assumption, more so even than the other two, begs the question. The purpose of Perrin's argument, according to the thing realist, was to establish that there are indeed things to underlie the random motions of the suspended particle in Brownian motion. To assume that there must indeed be such underliers is viciously circular.

4.4.3 Summary

It is easy to mistake Perrin's argument for an argument for the existence of things. However, we must remember that Perrin's argument is primarily an argument *against* the plenum view of matter. I.e., he was arguing that thermal systems are not composed of continuous substance. While it is natural to suppose that the diametric opposite of continuous substance is discrete substance—static things like atoms and molecules—it would be a mistake to conclude that Perrin's argument is naturally an argument for such things. In fact, Perrin draws attention to the true novelty of his arguments: that we substitute a "kinetic for the old static conception of the fluid state" (1916, 1). I.e., we are not meant to conclude that there are new things at all, but rather that what we thought was static is actually dynamic.

While the explanations of these dynamics offered by Perrin and others contain terms we typically associate with thing-like entities, the nature of these things is left entirely opaque. Or rather, the nature of these things is irrelevant. All that

---

[148] I note, again, that this argument that only things can be counted again begs the question against the process realist. If Perrin's argument is meant to establish the existence of things through the recognition of some granularity in our models, then surely the premise that only things exhibit granularity is not permissible. If things entail granularity, and granularity entails things, we are left in a dialectical circle.

matters is that they move and interact appropriately. Indeed, it is these motions and interactions that remain constant in multifarious models of thermal systems.

We should therefore expect the thing realist to provide an account of what makes these so-called underliers things at all. Such a task might be done by specifying those underliers as a physical system to be investigated, with the goal of producing thing-like features of the system. However, this was exactly how we began Chapter 3. The thing realist posited a thing-like entity. We then declared it a physical system to be investigated, and sought to explain its features. However, upon analysis, we discovered that we needed to posit that this system is a collection of dynamics because of the need to specify the system's features in terms of the experiments we perform on the system. If we but repeat this series of investigative steps, we should find that the new thing-system—atoms and molecules—are no more than collections of dynamics. This is indeed what we find (cf. Earley 2008a, b, c, 2012, 2016).

4.5 Conclusion

The thing realist faces a tension with historical analysis in interpreting Perrin. On the one hand, Perrin consistently refers to his program and arguments as a rejection of the "old statical" and "static" conceptions of fluids and other thermal systems. He thereby explicitly rejects the substance or thing-ontological paradigm. However, thing realists wish to interpret Perrin as arguing for the existence of a substantial and thing-like entity to act as the core compositional element of thermal systems. They must therefore attribute to Perrin arguments involving the thing-like nature of the mereological components of matter that would directly contradict Perrin's rejection of determinate, non-dynamic states of the fluid. If there are true things composing the thermal system, then the thermal system can be fully and independently described at a moment in time in terms of those things and their momentary, definite properties. However, this would mean that the thermal system has definite, independent, non-contextual, non-dynamic properties in virtue of its mereological nature, a claim that Perrin rejects.

This same tension is felt throughout the physics of the 20th century. Most notably the quantum physics that would succeed Perrin's matter physics in providing accounts of the fundamental material nature of physical systems. Once again, we see historical figures in quantum physics articulating explicit rejections of the substance or thing-ontological paradigm, while philosophers of science in the current day present us with interpretations of these thinkers that are explicitly thing-realist.

As but one example, consider the statements of Dirac to the Indian Science Congress in Baroda:

> When you ask what are electrons and protons, I ought to answer that this question is not a profitable one to ask and does not really have a meaning. The important thing about electrons and protons is not what they are but how they behave, how they move. I can describe the situation by comparing it to the game of chess. In chess, we have various chessmen, kings, knights, pawns, and so on. If you ask what a chessman is, the answer would be that it is a piece of wood, or a piece of ivory, or perhaps just a sign written on paper, anything whatever. It does not matter. Each chessman has a characteristic way of moving, and this is all that matters about it. The whole game of chess follows from this way of moving the various chessmen ... (Dirac 1955)

Once more, we see a physicist presenting us with an explicit reason to suppose that thing ontology is simply uninteresting. The things we posit are mere placeholders for particular motions, not definite, particular, non-contextual entities like the thing realist would wish to find. We do not make use of things in our explanations, nor is their ontological character important to the experiments or models of physics. How, then, can the thing realist reasonably seek to recover a thing ontology in the context of this physics?

This is not merely a philosophical tension, however. It has direct consequences in the interpretation of theory. One instance of this is the focus of the next chapter. Namely, I will show that the thing realist's ontological intuitions produce explicit contradiction in the interpretation of nuclear models. This means that thing realists must commit to something absurd if they wish to continue holding to these ontological beliefs. The process realist faces no such challenge, as I argue, and can in fact unify the supposedly incompatible nuclear models.

Chapter 5
Models of the Nucleus: Incompatible Things, Compatible Processes

5.1 Introduction

If nuclear models are taken to represent things—determinate, particular, independent, atemporal, entities like objects, structures, and substances—then they are incompatible (Boniolo et. al. 2002; Teller 2004; Morrison 2011; Portides 2011). Specifically, the two most prevalent models—the liquid-drop and shell models—treat the nucleus, its internal structure, and the component nucleons as entities with contradictory properties. These differences allow these two models to describe and explain different nuclear phenomena: fission and neutron scattering in the liquid-drop model, and single-nucleon excitation and nuclear decay in the shell model. However, prima facie, these differences also suggest that these models are incompatible in their ontology. Indeed, by maintaining an adherence to standard static ontologies of objects, structures, and substances, henceforth "thing realism,"[149] this incompatibility is irresolvable. Any success of their explanations comes not from the entities, properties, and structures they posit but from somewhere else.

This presents two problems. First, there is an ontological problem with this incompatibility: to what are we referring when we use the term "the nucleus"? Second, there is an explanatory problem: how can we reasonably use one set of features of the nucleus to explain successfully if those features are incompatible with equally explanatory features of a different model? The explanatory problem is built on top of the ontological problem. The ontological problem suggests that the explanations offered by, e.g., the liquid-drop model are reliant on entities that *cannot* exist, meaning that the explanations are akin to explaining the functioning of a computer with square circles. In other words, the incompatibility of these models is more than the result of attributing different features to a single system: each model denies the possibility of the (thing) ontology offered by the other.

However, this is only a problem with a thing interpretation of these models. By taking seriously the experimental methods by which these models are constructed

---

[149] Examples of thing realism abound. See Wiggns (2016a, b, c) for an example of a mixed ontology that treats substance as the continuant of an active principle.

and the calculational tools these models provide for interpreting experimental outcomes, I construct a new form of realism about these models that renders them ontologically compatible. Namely, I argue that nuclear models are consistent in the dynamic entities to which they refer: the interactions of nucleons, the excitations of the aggregate whole and its parts, the decay processes the system undergoes, and, in general, the processes of the nuclear system. Here "process" is a primitive term used to refer to the sort of entity exemplified by motions, interactions, excitations, growths, decays, etc.[150] Similar to recent work by other authors in the philosophy of science (Dupré 2010, 2018; Earley 2003, 2008a, c; Kaiser 2018; Guay and Pradeu (forthcoming); Guay and Sartenear 2018), I therefore advocate a pure process realism with respect to nuclear models. That is, we should reify the models' processes—the non-particular, dynamically contextual, uncountable, determinable entities—without reifying the objects, structures, and static properties that are demonstrably incompatible. By doing so, I argue that both the liquid-drop and shell models are fully compatible and successfully explanatory. The nucleus, therefore, is no more than a collection of processes.[151]

Critical to this process realism is the recognition that the processes referred to within nuclear models are essential parts of the observational acts that form nuclear experiments. In particular, because the dynamics within the nucleus must always be a continuous intermediary of our experimental interventions and the reception of signals from the system, these dynamics are essential dynamic parts of nuclear experiments. We are therefore licensed in inferring these dynamic parts on the basis of experimental practice alone because it is only experimental practice that makes true the descriptions of processes in nuclear models.

In contrast, the thing terms reified by the thing realist in these models require additional inferences, the premises of which cannot be supported on the basis of experiment alone. The essential premise of these additional inferences is one of two options (a) that all dynamics (metaphysically) require static things to underlie

---

[150] I use the word "process" in much the same way as Seibt (1990, 2004, 2005, 2018) uses the word "dynamics." Importantly, it is a primitive term and cannot be defined independently. See Chapter 1. See also Pemberton (2018), Griesemer (2018), Love (2018), and Chen (2018) for more on constructing a working definition of process within scientific theory (specifically biology).

[151] Similar claims have been made about other systems. Parr et. al. (2005) and Bader (1999, 2008) problematize the molecular system in a similar way. Hattema (2007) describes some of the problems in the history of molecular modeling, e.g., the Heitler and London (1927) and the Hund (1927) models of the molecule. Earley (2008a, b, c) argues explicitly for a "process structural realism" about the molecule that is quite similar to what I articulate here. The largest difference is that Earley is willing to reify structures within the molecule, while I take this as impossible given that structures in the nucleus are a part of the incompatibility of nuclear models.

them,[152] or (b) that the existence of stability in an experimental system necessitates something static and unchanging within the system. These premises are dubious if deductively supported, and insufficient if inductively supported. Thus, inferences to processes are experimentally supportable, whereas inferences to things are dubious at best. Process realism is therefore superior to thing realism in the context of nuclear models because it:
(1) establishes cross-model consistency,
(2) accords with experimental practice and explanations, and
(3) is ontologically and epistemically modest.

We begin with a brief introduction to the liquid-drop and the shell models of the nucleus (§5.2). In addition to the normal exegesis, I will also show how these models are ontologically incompatible with any form of thing realism (object, structural, substantial, bundle-theoretic, etc.). We then turn to reinterpreting the material and formal features of the nucleus in terms of dynamic entities (motions, excitations, etc.) (§5.3). In discussing the formal and material features meant to support the haeccity of the nucleus, I offer the standard process-realist arguments that these features are no more than real dynamics with the added knowledge that they cannot be reinterpreted as things in any sense. I then show that the explanations of the nuclear models rest on these dynamics, and that these dynamics are compatible. Their compatibility is the result of both models acknowledging the existence of all processes, but only using a subset of these processes in order to explain the salient behaviors of the nucleus.

5.2 Many Models, Divergent Things

There are many models of nuclear systems corresponding to many phenomena.[153] Liquid-drop models, developed in the 1930s, are still used to model the fission of a nucleus. Shell models, similar to molecular and atomic orbital models, are used to model nuclear line spectra, and to a lesser extent the stability of the nucleus. Lattice models are used to understand nucleus formation and structural binding sta-

---

[152] This is a claim that dates back to Aristotle. See *Physics* 190a31–b9. Seibt (1990) also criticizes this premise extensively.
[153] See Cook (2006) for an overview of the various types. One type not discussed by Cook, more recently developed, is the so-called unified lattice model. See Caurier et. al. (2005). It is worth noting that this model is not a unification of the liquid-drop and shell models. The hope is that the model will be able to unify explanations of nuclear decay and of scattering experiments, but the model cannot explain fission. It is still unclear that the model can explain spectral emission.

bilities. For our purposes here, we will consider only liquid-drop and shell models. Considering these two will be quite sufficient to establish that no robustness argument—an argument that terms in these models refer to real things in virtue of their appearance and similarity in both (and more) models—can be made for the reality of things in the nucleus, or for the nucleus qua thing itself. Insofar as "the nucleus," "nucleons," "Energy structure," etc. appear as identical terms in models of nuclear systems meant to refer to a static thing, the terms are wildly divergent in meaning. The models therefore contain no robust thing terms.

5.2.1 The Liquid-Drop Model

The liquid-drop model treats the nucleus as a drop of incompressible fluid of similar shape and structure to a drop of water. This analogy facilitates a highly accurate account of the nuclear binding energy and of how nucleons act together to produce collective, nucleus-wide motions such as fission. The model achieved its success primarily through this description of fission given by Meitner and Frisch (1939).[154]

In a drop of liquid, molecules will meaningfully interact only with their nearest-neighbors. For example, water molecules in a drop of water will electrostatically repel and attract each other and undergo collisions brought about by the momenta and thermodynamic vibrations of the molecules. Each of these interactions is between exactly two molecules, and the interactions are only significant when these molecules are sufficiently close to each other. The liquid-drop model of the nucleus imports near-exactly this reasoning to nucleon-nucleon interactions. The model assumes that there is a strong attractive potential, built from pairwise interactions between nearest-neighbor nucleons, binding the nucleus together, as well as electrostatic repulsion between nearest-neighbor nucleons that prevents collapse.

This assumption entails a difference in binding energy between particles at the surface and particles within the volume of the liquid-drop. Particles at the surface will always have fewer neighbors than particles within the volume. Thus, particles at the surface of the drop will be much more weakly bound than particles in the interior of the drop. This means that the binding energy of the liquid-drop is

---

[154] For a fully detailed explanation of the history of this model, see Stuewer (1994). See Gamow (1929) for the first liquid-drop model. See Cook (2006, ch. 4) for a detailed introduction to the liquid-drop model.

expressed by the following proportionality (k_1 and k_2 are real constants, mere proportions):

(Eq. 1) $E_{binding} \approx k_1(\text{number of particles}) - k_2(\text{number of particles})^{2/3}$

As the number of particles increases, the first term—the volume term—begins to dominate the second term—the surface term. Thus larger drops are less stable, since they have lower binding energy.[155] Following this analogy, the nucleus therefore exhibits similar behavior to that given by equation 1.

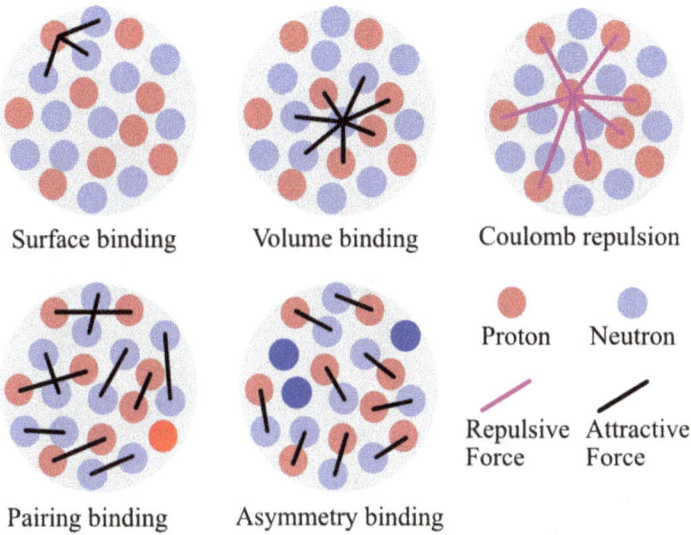

Figure 1: The nearest-neighbor interactions described by the liquid-drop model. Pairing and asymmetry binding forces are empirically, rather than theoretically, motivated.

Additional pairwise interactions are added to this nuclear model to provide further accuracy to the theoretically predicted binding energy (Figure 1). First, one recognizes that protons will repel each other. Therefore, an additional repulsive term is added to the proportionality of equation 1. Second, empirically motivated terms are added recognizing that nuclei with an even number of nucleons have higher binding energy (pairing), and recognizing that nuclei with an equal number of neutrons and protons tend to have higher binding energy (asymmetry). These are collected in the Weizsäcker mass-energy equation:

---

[155] Note that by convention, binding energy is considered large when it is a large negative number.

$$\text{(Eq. 2) } BE(Z, A) = k_1 A + k_2 A^{2/3} + k_3 Z(Z-1) + k_4 \frac{(A-2Z)^2}{A} + k_5 \frac{1}{A^{1/2}} + \text{shell correction terms}$$

Figure 2 shows the relative effects of each term in moving the theoretical binding energy curve closer to the observed binding energy curve.[156]

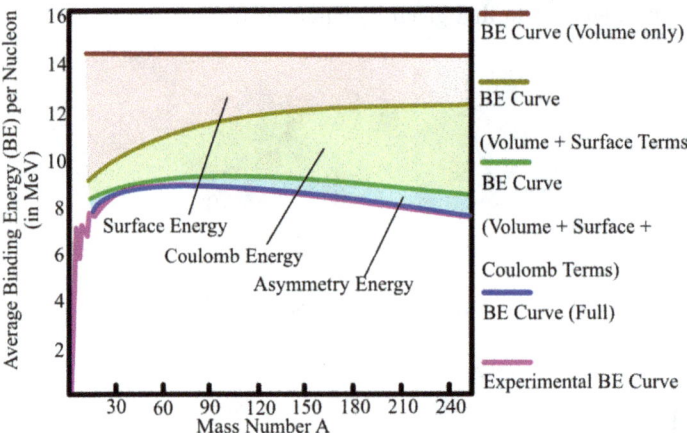

Figure 2: A depiction of the relative effects of each interaction type on the accuracy of the predicted binding energy curve. The energy of each interaction type is shown as the difference between curves (from Cook 2006, 61).

With all of these terms accounted for, the liquid-drop model is able to explain vibrational and oscillatory resonances of the whole nucleus. Any disturbance in the nucleus as a whole will be the result of many interactions between neighboring nucleons. For example, if a nucleus is struck by a low-energy bombarding neutron, this neutron's energy will distribute throughout the nucleus through the nearest-neighbor interactions depicted in Figure 1. Impacted nucleons will similarly interact, causing the nucleus as a whole to enter a higher energy state. The nucleus will redistribute the energy of the bombarding neutron among its nucleons until this energy is released through nucleon emission or fission.

---

[156] See Weizsäcker (1935) for the full mathematical treatment of these terms.

5.2.2 The Shell Model

While the liquid-drop model is concerned with analyzing the behavior of collections of particles, the shell model considers the behavior of an individual nucleon. The model seeks to explain only the behavior of this individual nucleon and treats all other nucleons as equivalent to an external Fermi field to which this nucleon couples. The nucleon is therefore treated as if it were part of a diffuse Fermi gas, much like the electrons bound in an atom. This facilitates a description of how an individual nucleon can occupy and transition between energy states within the nucleus. This in turn allows the model to explain the special stability of nuclei with certain numbers of nucleons: the so-called magic numbers.[157]

Two facts are suggestive of nuclear energy shell-structure. First, an individual nucleon will not collide with other nucleons very frequently. Were such collisions to occur, nuclei would be far less stable in various decay processes than observational data suggests. Second, for an individual nucleon, the forces acting on it from the other nucleons can be amalgamated into a single quantum potential well to which the nucleon couples—a Fermi field. Thus, the nucleon will enter quantized energy levels, the structure of which will depend on the shape of the potential well imposed on the individual nucleon. These facts entail that nucleons occupy non-coinciding trajectories within the nucleus. Given that nucleons experience strong attractive forces which would otherwise bring them into collisions, this in turn suggests that nucleon trajectories are kept apart by something like the Pauli exclusion principle. In analogy with the case of electron orbits in an atom, nucleons are unable to occupy the same trajectories, instead occupying discrete trajectories quantized by the total angular momentum and energy of the nucleon occupying that trajectory.

The applicability of the Pauli exclusion principle for nucleons facilitates a direct comparison of nuclear structure to atomic structure: the nucleus can be treated like an electron cloud in an atom. Electrons moving in an atom move in orbits—orbits that are separated from each other by Pauli exclusion—defined by a central, attractive potential well. Each electron occupies the lowest-energy orbital that is not already filled by another electron, and orbits are quantized in terms of the net energy, angular momentum, and spin of the occupying electron. Similarly, nucleons in the shell model occupy energy levels—"orbits" or "energy shells"—which are defined by something like the harmonic oscillator potential well. The energy levels associated with coupling the nucleon to such a harmonic potential well

---

[157] The first shell model is presented by Mayer and Jensen (1955). For a modern introduction to the model, see Cook (2006, ch. 2).

are depicted in Figure 3 (left hand lines). Two nucleons, like electrons in the atom, are incapable of occupying the same energy level in virtue of Pauli exclusion. Finally, having defined nucleon "orbits," one introduces a spin-orbit coupling for the orbiting nucleon. This splits the energy level into sublevels defined by the total angular momentum of the nucleon, as depicted in Figure 3 (right hand lines). Just like electrons in an atom, nucleons will occupy the lowest available energy level that is not already occupied by another nucleon, and nucleons will transition between two levels only when individually excited with the discrete energy corresponding to the difference in energy between the two levels.

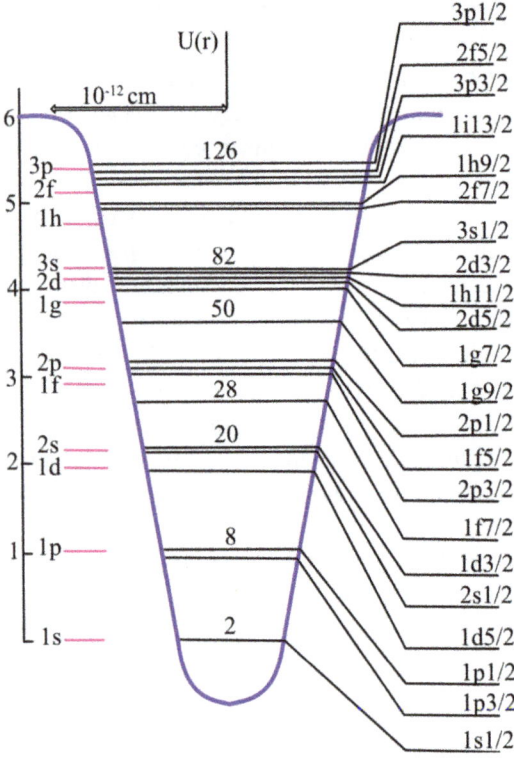

Figure 3: Shell quantization in the shell model, showing the energy-momentum shells (left) and spin-orbit subshells (right). Shell closure occurs at the indicated magic numbers of nucleons (middle): 2, 8, 20, 28, 50, 82, 126, at which the nucleus is most stable (from Cook 2006, 75).

This provides the basis for explaining the magic numbers. Atoms with filled energy shells exhibit much greater stability than atoms with open energy shells, called shell closure. This is why noble gas elements like Helium and Neon are far less reactive than elements like those in the alkali group, e.g., Lithium and Sodium. Helium and Neon have "magic numbers" of electrons—2 and 10—which correspond to the number of electrons needed to fill the lowest energy levels. Similarly, nuclei which fill all the subshells in a given energy shell will experience shell closure, and be much more stable. Shell closure occurs when a nucleus has 2, 8, 20, etc. nucleons of either type.

5.2.3 Incompatibility

We now ask: are these two models describing the same thing? The answer is that they are not. The nucleus of the liquid-drop model is quite different from that of the shell model. The two models of "the nucleus" describe the nucleus as having a different shape, different internal structure (e.g., definable spatial relations between nucleons), different spatial extent and density, and different analogous material constitution. This is summarized in Table 2.

Table 2: A comparison of properties and thing-claims made by the liquid-drop and shell models

Feature of the Nucleus	Liquid-drop Model Claims	Shell Model Claims	Experimental Results Support
Entity Claim	The nucleus is a liquid-drop.	The nucleus is a Fermi gas.	Neither
Presence of Internal Structure	There is no internal structure.	There is internal structure.	Both
Shape of the Nucleus	Roughly spherical, varying directly with nucleon number	Roughly spherical, not varying directly with nucleon number	Liquid-drop
Density of the Nucleus	Constant, with a sharp drop at the nuclear radius	Varying, with a gradual drop at the nuclear radius	Shell
Radius of the Nucleus	Proportional to $A^{1/3}$ (the cube root of the number of nucleons)	Proportional to occupation numbers of energy shells, not $A^{1/3}$	Liquid-drop

First, the liquid-drop and shell models directly contradict each other on the presence of internal nuclear structure. The liquid-drop model admits only a nuclear surface and interior. In contrast, shell models predict energy shells and subshells, and therefore admit rich internal substructure. Thus the liquid-drop model depicts a nucleus with *no* internal structure, and the shell model depicts a nucleus with *much* internal structure.

The two models also conflict over the size and shape of the nucleus. The liquid-drop predicts an approximately spherical shape for the nucleus resulting from nearest-neighbor attractive interactions between the nucleons. This in turn means that the liquid-drop predicts a sharp drop in nucleon density as one approaches the boundary of the nucleus. There are few-to-no nucleons that exist outside of the nuclear radius. The nucleus, according to the liquid-drop model, has a definite shape and size related to the number of nucleons A; the nucleus is a spheroid of radius proportional to $A^{1/3}$.

In contrast, the shell model predicts that the size and shape of the nucleus depends on the shape and closure of energy shells. This is because nuclear structure in the shell model is almost entirely dependent on the texture of these shells. The nuclear radius will therefore depend on occupation number, not $A^{1/3}$. In addition, the density of nucleons will vary radically between nuclei with magic numbers of nucleons and nuclei without magic numbers of nuclei, again because energy shell structure determines nuclear structure. These features are depicted schematically in Figure 4.

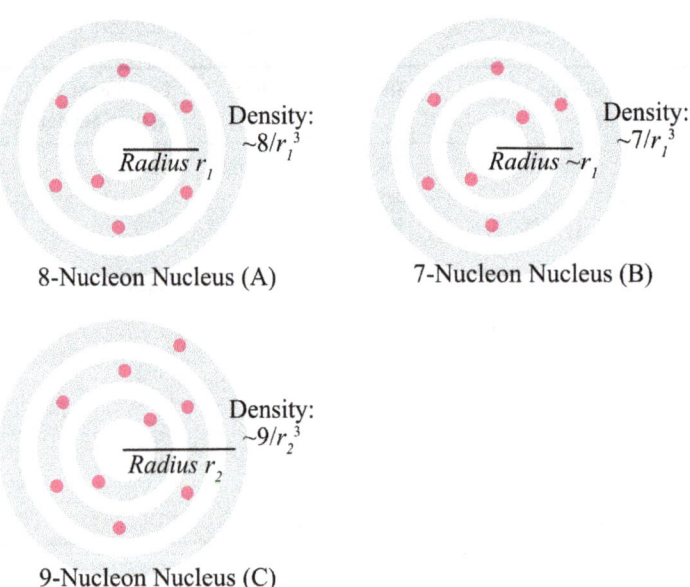

8-Nucleon Nucleus (A)   7-Nucleon Nucleus (B)

9-Nucleon Nucleus (C)

Figure 4: A representation of nuclear density and radius as related to nucleon number. Large differences in both properties are observed between A and C, and B and C.

Finally, the energy shell structure of the nucleus also entails that nuclear density varies continuously with increasing radius. Rather than a constant density as in the liquid-drop model, the shell model predicts that the tendency of energy shells to spatially separate will cause nucleon density to be inconstant, especially in nuclei with open energy shells.

Empirical data do not strictly rule in favor of either model. Rather, both models experience some success in their explanations, and failure in others. The liquid-drop model successfully explains the nuclear radius but not the nuclear density. Fission experiments and neutron bombardment indicate that there is little internal shell structure in the nucleus, supporting the liquid-drop, while scattering experiments and observations of radioactive decay suggest the rich internal structure of the shell model. The models are therefore directly contradictory in a thing ontology, with no apparent way of adjudicating between them. It is no wonder, then, that many authors advocate not only that these models are incompatible, but that we should remain silent on, or else eliminate, thing-realist intuitions about the nucleus on the basis of these models (Boniolo et. al. 2002; Teller 2004; Morrison 2011; Portides 2011).[158]

5.3 The Features of the Nucleus Are Processes

The liquid-drop and shell models represent contradictory things. Therefore, these things cannot be robust. However, we saw in Chapter 2 that this is not special. Indeed, we expect that processes are robust where things are not. To see this in the nuclear case, we return to our list of features of the nucleus. These are shown in Table 3.

Table 3: A list of formal, material, and productive features of the nucleus to be explained

	Formal Features	**Material Features**	**Productive Features**
Example Features	– The shape of the nucleus – The internal energy spectrum structure – The radius of the nucleus – The density of the nucleus	– The composition of the nucleus in terms of protons and neutrons – The composition of decay products produced by nuclear radioactive decay	– The production of line spectra – Decay product production – Fission – Fusion

---

[158] Cook (2006) also remarks that the models make inconsistent claims. See also Bohr and Mottelson (1969) for an in-depth, historical account of the tension between these models.

Importantly, the features of the nucleus that are of most interest are the formal and productive features. These are, at minimum, the features of the nuclear system that models of the nucleus are supposed to explain. We saw this in section 5.2: the liquid-drop model can successfully explain the shape and size of the nucleus, and the shell model can successfully explain its internal energy structure.

However, there is an important distinction we must draw: the formal features are not, strictly speaking, the explananda of nuclear models. The liquid-drop and shell models are not used in experimental settings, both historical and contemporary, to explain these formal features. Rather, they are used to explain the processual features of the nuclear system: fission, radioactive decay, spectral emission, neutron capture and scattering, etc. These are the phenomena that are actually modeled and occur in experimental settings. The formal features of the nuclear system are therefore explananda only insofar as we might already have thing-realist intuitions.

Moreover, I have already shown how the formal features of the nucleus according to the two models are incompatible. Therefore, these formal features cannot be explanans either. Rather, I show below that the formal features of the nucleus are placeholder terms for collections of processes, useful only for their pragmatic role in describing the evolution and dynamics of interest in the nuclear system. Importantly, this is not an a priori argument, but rather follows simply from the facts of the models and their history. The models were designed to explain dynamics, and use dynamics to do this explaining. This is the explanatory defeat of things: processes do all of the explaining in these models, and are the entities being explained.

In turn, the material features are offered as the thing realist's hope of an explanans independent of these formal features. Surely, the argument goes, the material composition of the nucleus plays a role in explaining nuclear phenomena simply because nucleons are the bearers of properties and vehicles of processes in the nucleus. As I will articulate, this argument, just like the underlier arguments rejected in Chapter 2, fails because it does not rule out that the nucleons are themselves processes or collections thereof.[159]

However, just like with the formal features, the incompatibility of nuclear models on thing-realist interpretations makes it difficult to see how appeal to

---

[159] Note that, just as I argued in Chapter 2, this means both that processes can act as truth-makers for sentences involving change and persistence (cf. Stout 2012), and that where necessary, systems of processes act as a "persistent entity" to ground claims about, e.g., the nucleons in nucleon-nucleon interactions and the like. This latter claim, that systems of processes can ground structural properties of a physical system, entails the former claim.

thing-nucleons is meant to resolve anything. Given that these nucleons have to be fit into incompatible models as explanatory elements, they will inevitably inherit that incompatibility. For example, if nucleons are meant to both occupy discrete energy shells and not occupy these shells, the nucleons themselves will need to have a property[160] "energy within the nuclear system" that is one value in the liquid-drop model and another in the shell model.

This means that things are defeated ontologically as well. No thing term may be reified in both of these models on pain of contradiction, and no thing term may be thought to explain in either of these models. Instead, as I show below, all we need are processes. Processes are explanans, explananda, and ontological ground in both models. Moreover, the processes of each model are compatible with each other. I turn now to a re-analysis of the features of the nuclear system, in order to show this.

5.3.1 Formal Features of the Nucleus: Balanced Dynamics

For the sake of simplicity, let us consider only two formal features of the nucleus: the shape of the nucleus and the internal structure of its energy spectrum. We use the liquid-drop model to explain the former, whereas we use the shell model to explain the latter. Moreover, the two models provide correct explanations for their respective features of the nuclear system, but fail to explain the other feature. Naïvely, this suggests that, even independently of the incompatibility of the thing interpretation of the two models, these two models cannot find common ground. As we will see, the processes used to explain these two features are indeed common ground between the models: the models do not contradict in their process-realist interpretation.

5.3.1.1 The Shape of the Nucleus
The shape of the nucleus is explained by the liquid-drop model. This explanation is simple: the shape of the nucleus is the result of all of the inter-nucleon interactions (as cataloged by the Weizsäcker mass-energy equation) being balanced against each other to minimize the total energy of the nuclear system. In other words, the electromagnetic and chromodynamic interactions, both represented as simple modifiers on the strength of each term in the Weizsäcker mass-energy equation, counteract each other to bring about a stable configuration of the system. In

---

[160] Note that the property need not be basic or fundamental, but will need to be constructible in order for the thing realist to reify the nucleon.

other words, the shape of the nucleus is not some static feature of the nucleus, but is rather the result of a balance of multiple dynamics within the system. This matches other discussions of structures in chemistry (Earley 2003, 2008a, b, c) and physics (Finkelstein 1996, 2008; Chapter 3).

However, it is far more interesting to note how the shape of the nucleus plays a role in the explanation of fission. Data on neutron bombardment of nuclei puzzled physicists in the 1930s (Stuewer 1994). The best theories of the nucleus predicted that for low-energy neutrons, neutron capture should be as likely as scattering from the nucleus. However, experimental results showed that neutron capture was much more likely. In addition, data showed that for each element, neutrons with certain energies were absorbed at higher rates. This too disagreed with the current theories of the nucleus.[161]

Bohr's compound nucleus provided the solution to these problems. Bohr analyzed the process of neutron bombardment into several distinct stages. First, an incident neutron impacts the nucleus. Second, the nucleus absorbs the neutron, and the neutron's energy is distributed among the nucleons in nearest-neighbor interactions. Then, if enough energy is collected into a single nucleon, that nucleon is ejected from the nucleus.[162] However, if there is not enough energy to eject a single nucleon, the nucleus captures the incident neutron. The energy required to eject neutrons from a nucleus of a given element depends upon the particular binding energy of that nucleus. Thus, only neutrons with particular energies will be able to "scatter" by being ejected from the nucleus: a compound nucleus with captured neutron is more likely to form if the energy of the incident neutron is enough to excite the nucleus into the next highest energy level of the nucleus as a whole. This effect, which Bohr (1936, 344) called "resonance excitation" explains why neutron capture is more likely than scattering, and why certain neutron energies produce peak capture or peak scattering. Here, resonance is taken literally, unlike in discussions of molecular resonance. Bohr is literally describing the nucleus as resonating, i. e., vibrating with harmonics in tune with the incident energy of the bombarding neutron. When resonant, the nucleus enters a standing wave oscillation pattern. It is these standing wave oscillation patterns that constitute the energy levels of the nuclear system, just like the standing waves of a string on a cello.

---

[161] See Pais (1981, 336–337) for more on this historical difficulty.
[162] See Bohr (1937, 163) for this exact account, with explicit reference to both processes and stages (intermediate states) of a process. See also Bohr (1936).

This explanation was later extended to include the emission of larger clusters of nucleons by Meitner and Frisch (1939).[163] Instead of emitting only a single nucleon when in an excited state, Meitner and Frisch proposed that oscillations of a compound nucleus split that nucleus into two smaller nuclei through the same process as Bohr describes. This is depicted schematically below, in Figure 5.

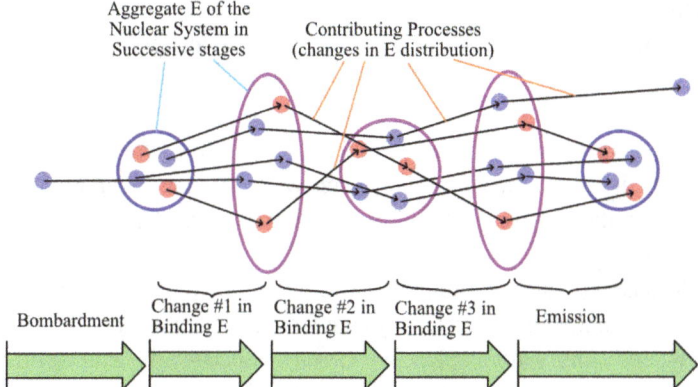

Figure 5: A schematic of the identifiable processes involved in an instance of nuclear scattering event as described by the liquid-drop model.

These oscillations of the nuclear system are oscillations in the shape of the nucleus. However, these oscillations are the result of a disturbance of the balance of strong and electromagnetic interactions. When we bombard the nucleus with a neutron (our intervention), this bombardment produces a change in the system: the emission of two or more fission products. We observe that the fission products have a characteristic kinetic energy. We therefore infer that this kinetic energy—the motion of the fission products—must have been acquired through some redistribution process within the original nuclear system. We therefore infer, following Bohr, that this redistribution of motion within the nuclear system is the result of a series of interactions, first between the nuclear system and the bombarding neutron, then between nearest-neighbor nucleons. We then represent this redistribution as a collective motion of the nuclear system, effectively averaging over the many individual interactions to produce the oscillations of the nuclear system. It is at this point that we represent this collective motion as an oscillation in the shape of the nucleus. Such a representation is good because the shape of the nu-

---

[163] See Stuewer (1994, 107–116) for a full account of the historical development of the liquid-drop account of fission offered by Meitner and Frisch. See also Frisch (1939) and Meitner (1936).

cleus acts as an effective placeholder for the balance of processes within the nuclear system. E.g., when the nucleus has a "spherical" shape, the processes are perfectly balanced and therefore the system is stable. However, when the nucleus has an "oblong" shape, the processes are *imbalanced*, and the system is unstable. By describing fission as the result of oscillations in this shape from spherical to oblong, we thereby show how the nuclear system dynamically reaches the point of instability at which fission occurs.

Crucially, the shape of the nucleus is acting as a placeholder term in this explanation. Namely, it is a placeholder for the balance of many sub-nuclear interactions and motions. When we want to describe these interactions and motions in aggregate, shape becomes a relevant feature of the system. However, "shape" is only relevant when it itself is dynamic. I.e., even when we are making use of "shape" in our explanations, we are implicitly referring to the underlying interactions and motions of the nuclear system.

Moreover, the shape is only ever inferred, never observed. Rather, what we observe is the dynamic change of the nuclear system from stable to unstable that results from our dynamic intervention. Since we *already* associated the shape of the system with the stability produced by balanced internal interactions, we describe this dynamic change from a stable to unstable nuclear system in terms of the shape of the system.

5.3.1.2 Internal Energy Structure

We turn next to the explanation of the internal energy structure of the nucleus provided by the shell model. Again, this explanation is rather simple, and quite obviously processual: the energy structure of the nucleus is the result of the interaction of single nucleons with the aggregate potential created by the remainder of the nuclear system. This interaction can be further divided into an electromagnetic, chromodynamic, and spin interaction. Thus, the available energies of a nucleon in the nucleus are defined by these three interactions that nucleon can have with the nuclear potential. It should come as no surprise, at this point, that the bound state energies that these interactions produce are the result of balancing these three interactions. I.e., the relative strengths of the three interactions define a series of stable states in which the three interactions are balanced. There being multiple ways to balance these interactions, there are correspondingly many ways in which the nucleon can occupy a stable energy level.

Again, it is far more interesting to see how and why this energy structure appears as a feature of the nucleus in the first place. As already noted, the shell model of the nucleus is motivated by a qualitative analysis of patterns of naturally occurring stable isotopes, their spectral line signatures, and the decay of non-magic-number isotopes into magic-number isotopes. Quantitatively, the stability of magic-number nuclei is the result of a difference between the binding energies of isotopes with a magic number of nucleons and nuclei with one additional proton or neutron.

This energy difference manifests in various observed decay and spectral emission processes. Lighter nuclei with one more nucleon than a magic number tend to decay through nucleon emission and alpha decay (Mayer and Jensen 1955, 21). For example, Helium-5 and Lithium-5 will both α-decay into an α-particle and the additional neutron or proton respectively. Heavier nuclei will instead tend to β-decay into more stable isotopes.

Mathematically, we represent these decay and spectral emission processes as the result of changes in the energy of individual nucleons. These changes are in turn represented by jumps between energy levels defined by the interaction of the nucleon with the chromo-electrodynamic potential created by the rest of the nuclear system. I.e., nucleons jump between energy levels quantized according to Figure 3 above.

This means that the energy structure is apparently playing an explanatory role in our explanations of decay and spectral emission processes. However, just as with the shape of the nucleus, the energy structure of the nuclear system is only acting as a means of identifying the relevant processes in the system. Recall that the goal of the shell model is to explain decay and spectral emission processes. It was for this reason that we constructed energy states. We first intervene on the nuclear system by, e.g., bombarding it with light. This produces a change in the system, namely the emission of line spectra. We then infer that our intervention must have produced, through a series of dynamics within the system, this emission of line spectra. We infer that it is single nucleons that are excited by this light, and which, in losing this excitation energy, emit the line spectra we observe. We know from the frequencies of the line spectra the energy of each emission, and therefore the energy of each energy excitation in a single nucleon. We therefore *construct* a collection of energy states that can exhibit these energy excitations, i.e., by ensuring that the energy of each possible excitation is equal to the difference in energy between two energy states. Importantly, we *choose* the mathematical potential in which we define these energy states. While we know the interactions that define this potential (chromodynamic, electromagnetic, and spin interactions), we do not know their relative strengths or balance within the nuclear system. This is something we must do entirely based on fitting our model to the observed energies of spectral emission (and decay).

In short, the energy structure is superfluous to our explanations of decay and spectral emission. Our interest is only in the balance of interactions between nucleons, and the interaction of, e.g., radiation with this balance that produces the absorption, excitation, and spectral emission processes we observe. I.e., not only do we define the energy structure of the nuclear system in terms of three interactions and their balance, we only ever make use of this energy structure in our explanations as a pragmatic means of identifying the relevant and explanatory processes.

5.3.1.3 Robust Processes in Formal Feature Explanations

Manifestly, both the liquid-drop and shell models of the nucleus contain reference to exactly the same processes: chromodynamic interaction, an attractive force, and electromagnetic interaction, a repulsive force. Thus, manifestly, these two interactions are robust features of the models. If we only reify these two interactions, i.e., the processes in each model, the models will not be contradictory.

Each model constructs its token, explanatory processes out of these basic interactions. For example, in the liquid-drop model, the difference between surface and volume chromodynamic interactions is the result of the difference in number of nearest-neighbor chromodynamic interactions that take place at the surface and within the volume of the system. This difference gives rise to two sorts of aggregate interaction, the surface-attractive interaction and the volume-attractive interaction. Similarly, the field interactions for individual nucleons within the shell model are the result of aggregating all of the chromodynamic and electromagnetic interactions of the other nucleons with the one in which we are interested.

Nevertheless, one might still be troubled that the two models disagree on empirical results. Even if they both refer to the same processes, they cannot both be correct in every empirical context. My contention, and what I show below, is that the difference between the models rests on their emphasis on how many and which processes to consider relevant, and which ones can be neglected in specific empirical instances. Neither model denies or contradicts the existence of the neglected processes. Quite the opposite: both models include explicit reference to the processes they are neglecting. However, both neglect certain processes precisely because they are not always relevant to the behavior being explained by the model. This being the case, I show here that the models do indeed become fully compatible precisely because neither is interested in being universally applicable.

The key to understanding this is that the liquid-drop and shell models do not posit processes in a vacuum, without context.[164] Rather, the processes of both models only exist insofar as they are connected to specific interventions and dynamic alterations of the system. Thus, the processes posited by each model, themselves

---

164 See Jungerman (2003) for a brief discussion of the importance of interconnectedness both in physical models and process ontology. Process philosophy, especially as found in Daoist, Zen Buddhist, or Kyoto school traditions, relies on the contextuality of processes.

not directly observable as is, e.g., the emission of spectral lines, are contextualized to these interventions and dynamic alterations.

Let us consider again the shape of the nucleus. My contention here is that the nucleus only has a shape insofar as we interact with the nucleus in a particular way. In performing scattering experiments, we discover that there is a typical deflection pattern of our scattering probes. In collecting this information through many scattering trials, we can summarize it by claiming that the nucleus has a contained charge density that is responsible for the scattering. I.e., the nucleus has a shape. However, this shape is no more than a representation of the many individual deflections of scattering probes, the many interactions of the probe with the nuclear system. These deflections, then, do not involve the processes of, e.g., spin interaction described by the shell model. Instead, the interactions of the nucleons described by the liquid-drop model are the processes that are relevant to the scattering and subsequent deflection processes.

The processes of the liquid-drop model therefore do not deny those of the shell model. Rather, the liquid-drop model recognizes that those processes are not relevant to the processes that define, and in which we determine, the shape of the nucleus. In the process of fission, and the processes of energy redistribution that make up the intermediate dynamics of fission, only nearest-neighbor interactions are relevant. Interactions between distant nucleons still occur. However, they are of a size that is negligible in the redistribution of energy within the nucleus. This is because the redistribution of energy in the nucleus fundamentally relies on single nucleons translating their energy to others directly (hence the term "collision" in Bohr's explanations). In contrast, distant-nucleon interactions become highly relevant when considering the excitation processes of a single nucleon. For this reason, the shell model includes these distant-nucleon interactions as important features. Importantly, the shell model does not treat these distant-nucleon interactions as any different in strength or variety from the ones that the liquid-drop model neglects. They are the same processes in both models. The difference is that the shell model considers them relevant parts of the dynamics of decay and spectral emission, whereas the liquid-drop model recognizes them as small enough to neglect for the purposes of mathematical simplicity.[165]

---

[165] Notice that as the number of nucleons increases, the importance of the asymmetry and pairing terms in the Weizsäcker equation becomes more important (see Figure 2, the asymmetry energy). Something similar happens with the shell-correction terms in this equation. Thus, for larger nuclei, it becomes relevant to reincorporate the non-nearest-neighbor interactions into the model in order to explain nuclear shape. This is a further indication that the liquid-drop model is not contradicting the shell model in its process terms. Far from it; the liquid-drop model is affirming the dynamic ontology of the shell model.

This, then, is how we explain the difference in empirical predictions between the two models. The two models have different explanatory aims. Namely, they are attempting to describe and explain (in terms of specific processes!) different dynamic phenomena. This difference in aim translates into the incorporation of different processes as relevant aspects of the intermediate dynamics of the nuclear system. Spectral emission, nuclear decay, and fission simply involve different sequences of dynamics, just as we learn about each through different interventions and different acts of observation. Nevertheless, both models consider the nuclear system to be composed of electromagnetic and chromodynamic interactions, and of all the multifarious combinations of these basic interactions that one could reasonably construct. Thus, process realism about the liquid-drop and shell models produces a consistent, monist, realist interpretation of the models.

This is the explanatory defeat of things. Notice that the formal features of Table 2 are exactly those features of the nuclear system that the thing realist would expect us to explain using the liquid-drop and shell models. These explanations, as I have argued, are performed entirely in terms of processes: it is the processes of chromodynamic and electromagnetic interaction that explain the shape of the nucleus and the internal energy structure of the nucleus. In contrast, the supposed material features of the nucleus are those entities the thing realist would have us reify in order to explain the formal features. In short, the thing realist must argue that the explanations of formal features of the nucleus (and more importantly, the dynamics of fission and spectral emission and the like) require these material features in order to underlie the processes that are actually doing the explanatory work. I.e., the thing realist must offer an underlier argument that processes require things *ontologically*.[166]

5.3.2 Material Features of the Nucleus

We turn now to material features of the nucleus: the composition of the nucleus in terms of nucleons. Prima facie, the liquid-drop and shell models of the nucleus agree on these features. Both models, after all, refer to nucleons in their explanations of nuclear behaviors. So, prima facie, the thing realist seems to have an available interpretational move: the nucleus is a system composed of smaller things. While the formal features of the nucleus are (collections of) processes, are explained by processes, and are used in the models to explain processes, the nucle-

---

[166] The first explicit underlier argument was probably offered by Aristotle in *Physics* 1.2, 184b15–16.

ons may yet act as some sort of thing underlier for nuclear processes. The contention is both an ontological and explanatory claim:

(*Ontic Underlier Claim*): Processes cannot exist without vehicles to undergo them. For example, a nuclear interaction presupposes that there are two things interacting.

(*Explanatory Underlier Claim*): Processes cannot explain without things. For example, any explanation involving a change in the nuclear system, like nuclear resonance oscillations, requires that there is a thing that changes in order to explain the change itself.

There are three problems with this move, discussed in turn.

5.3.2.1 Nucleons Qua Things Do No Explanatory Work

First we have already seen that it is the nuclear processes that are doing explanatory work. Therefore, nucleons are only explanatory insofar as they participate in these processes: it is only the dynamics of nucleons that are explanatory. To suppose further that these dynamics deserve their own explanations is legitimate. However, it is illegitimate to suppose that these further explanations must involve things. The nucleon, just like the nucleus, is an experimental system with features to be explained, in particular a set of dynamics to be explained. Just like the explanations of nuclear features and behaviors above, we should suppose that the dynamics of nucleons, the behaviors of these nucleon systems, will be (and are in fact) explained by further processes, not things.[167]

5.3.2.2 The Underliers of Nucleon Dynamics Are Processes, Not Things

The thing realist must therefore rely on the ontological underlier claim. This leads us to our second problem. Specifically, any argument that is meant to support the existence of a thing-nucleon is equally an argument for the existence of a process-nucleon. I.e., any argument that there is a thing, "the nucleon," to underlie nuclear processes fails, or fails to rule out that this underlier is itself a process. This failure

---

[167] I will not discuss this further, since it is beyond the scope of this work. However, notice that nucleon dynamics and formal properties were explained from the beginning in terms of exchange interactions. This is the heart of Heisenberg's 1932(a, b, c) work on the nucleus (see especially 1932a). While Heisenberg's theoretical framework was rejected, the idea of nucleons being fluctuations of oscillations in the energy of an exchange force was taken up by Yukawa (1935) in order to describe cosmic rays, and later reintegrated into the shell model of the nucleus. Moreover, the nucleons are now described field theoretically as the balanced and stable interaction of triplets of quarks, which are in turn described as fluctuations in the energy of a quark field.

is the result of a general failure of such arguments: they rest on first recognizing that there is some stability in a system, and then claiming that this stability entails the existence of something static.

In the case of the nucleon, we do not need any general failure of underlier arguments to see that the nucleon is itself a collection of further processes. Nucleons are at best atypical things. Our nuclear models, as well as models of these particles in field theory, already suggest that these supposed things do not possess characteristics like any thing we have discussed so far. Nucleons are non-local, non-localizable entities in quantum field theory. They appear as fluctuations within an infinite field. Mathematically, they appear as creation and annihilation operators, mathematical entities that are typically associated with actions performed on and activities of a system, not objects within the system.

To see this, let us take the most basic feature of nucleons and see if they bear thing-realist interpretation. That is, consider the simple and basic property that nucleons come in two forms: protons and neutrons. These two things play quite similar roles in both the liquid-drop and shell models of the nucleus. Both engage in the same excitation processes, both engage in "collisions" that redistribute kinetic energy throughout the nucleus, both engage in collective motions as in fission, etc. In light of this, one may argue that protons and neutrons are not differentiated by their behaviors (their dynamics, the processes they undergo). Rather, one might suggest that the haeccity of protons and neutrons, or at least their "essential" properties of charge, mass, and spin, are the means by which we differentiate them. I.e., the thing realist might argue that neutrons and protons undergo the same processes (spin, excitation, collision, etc.), and yet are different entities. What differentiates them, then, must be non-processual. Hence, protons and neutrons, in virtue of being identifiably different, are things, not processes.

The point is well made, and is tempting. However, we cannot accept this line. Protons and neutrons do not undergo the same processes. Undoubtedly, many of the processes we associate with neutrons in the nucleus are similar to those processes of protons. However, we also always associate with protons an electromagnetic interaction of a strength on the order of the electron charge that we do not associate with neutrons. The proton and the neutron undergo different internal chromodynamics—different quark interactions. In turn, these quarks undergo different Higgs interactions with different relative strengths. These different interaction strengths are used as the definition of the different quark masses, and by extension, the different masses of the proton and neutron. Protons and neutrons undergo different decay processes, even within larger-scale models of the nucleus like the shell model. Neutrons can beta decay into a proton, an electron antineutrino, and an electron. Protons, however, beta decay into a neutron, a positron, and an electron neutrino. Protons and neutrons are therefore identified (and differentiated) both by different processes and similar processes of different strengths. They are therefore differentiated, by definition, in mathematical description, and by experiment, through differences of dynamics, not by determinate properties.[168]

This leads us to our first counterclaim against the ontic underlier claim: processes, not things, underlie processes. For any nuclear processes for which the thing realist supposes that we should include a thing underlier, "the nucleon," we need only replace the thing term with the relevant process underlier. For example, the underliers of nucleon-nucleon interaction in the liquid-drop model are the more stable quark-quark interactions that define neutrons and protons. The underliers of beta decay processes of the nucleus in the shell model are the two different beta decay processes associated with independent neutrons and protons. And so on. Provided the processual underlier is more stable than the process it is meant to underlie, the processual underlier is perfectly capable of acting as an underlier. Things are unnecessary.

5.3.2.3 Things Inherit Incompatibility

However, the situation for thing-nucleons is far worse than simple impotence. In fact, if nucleons are treated as things with thing-like properties, they will inevitably inherit all of the incompatibility of the models in which they appear. This means that, even were we to suppose that the nucleus is composed of things (nucleons), we would still be left in the position in which we began our discussion: with two sets of incompatible thing-components suited for successful explanations of physical phenomena in different contexts.

To see how nucleons inherit incompatibility from nuclear models, we need look no further than the claims of energy structure within the nucleus. The liquid-drop model consists of no internal energy structure. The shell model consists of rich internal energy structure. Now, supposing that nucleons are things with static properties and relations to other things, we notice that nucleons in the shell model must bear relations to other nucleons that are not born by the nucleons of the liquid-drop model. In particular, shell-model-nucleons must bear the relations that compose the differences in energy of the various shells. This means that the nucleons of the liquid-drop model, which do not bear these relations, are incompatible with the nucleons of the shell model. This same pattern can be

---

[168] Notice that the various determinate properties of the nucleons are only defined in physical models in a dynamic context. I.e., it is an essential feature of the quantum mechanical paradigm that the "properties of a nucleon" are determinables, not determinants, that are determined through a specific action on a system.

demonstrated for properties like "being a surface nucleon," "having a nuclear energy state," and other monadic properties, in addition to relations.

We might suppose that nucleons are not defined essentially in terms of these incompatible relations or monadic properties. This at least removes the incompatibility. However, if nucleons are not meant to be the bearers of properties and the relata for nuclear relations, then surely they are utterly impotent even within thing realism. Their only purpose is to satisfy some a priori assumption that the nucleus is composed of things. Such an assumption could hardly prove particularly informative, useful, or persuasive.

5.3.3 Reestablishing Compatibility

We can now collect what we have discussed in Table 3 (similar to Table 1 showing the incompatibility of thing interpretations of the liquid-drop and shell models) to show that the models are compatible on a pure process ontology (Table 4).

Table 4: A collection of the processes that appear in both models. These collections are identical, but are used in different ways to explain different ends.

Model →	Liquid-Drop	Shell
Explanandum of the Model	Fission, Capturing, and Scattering Processes	Excitation/Decay Processes
Explanans Offered by the Model	Aggregate motions of the nucleons, brought about by the nearest-neighbor interactions of the nucleons	Interaction of a single nucleon with the field of aggregated interactions provided by all other nucleons (Shell Interactions)
What Processes Compose the Explanans	– Chromodynamic interactions – Electromagnetic interactions	– Chromodynamic interactions – Electromagnetic interactions

Moreover, we can demonstrate this compatibility by locating in both models the processes contained in the other. As I have already discussed, the liquid-drop model contains explicit reference to shell interactions in the Weizsäcker mass-energy equation in the form of shell correction terms. In the majority of cases, these correction terms are unnecessary to include. Their effect on the binding energy of the nucleus, and therefore on the aggregate motions of the nucleus, is miniscule compared to the effect of nearest-neighbor interactions. However, for particularly large or small nuclei the shell correction terms become relevant. For small nuclei, this is because the shell interactions are on the same length scale as the nearest-neighbor interactions. Nearest-neighbor interactions are therefore significantly impacted by the small size and occupancy constraints on nucleons in their lower energy shells. For larger nuclei, this is because the source of shell interactions—the aggregation of all nucleons save one into a single Fermi field—is strong enough to noticeably perturb the nearest-neighbor interactions that guide fission and other aggregate motions. Thus, while the liquid-drop model has little to say about the form that these shell interactions take, the model still assumes that these interactions are occurring, and includes them as essential explanatory features when they become relevant to the modeler.

The same can be said of the shell model and nearest-neighbor interactions. The shell model is predicated on the fact that all nucleons in the nucleus are affecting the nucleon of interest. The potential well formed by the sum of all of these interactions is what defines the decay and excitation processes for a given nucleon by defining the available energy levels for that nucleon, its shells. As these interactions change, the experimenter making use of the shell model will need to alter the potential well in order to reflect these changes. In other words, the potential well is fine-tuned to reflect the specific nucleon-nucleon interactions, both nearest-neighbor and other, so that the experimenter may treat these individual interactions in aggregate.

In essence, both models refer directly to the same processes—chromodynamic and electrodynamic interactions—in their construction. Their difference lies in that they differently aggregate and select the processes that are relevant to their respective explanatory goals. While the shell model does not explain fission, it was never designed to do so. Since the shell model nowhere denies or contradicts the existence of those processes that *do* explain fission, we avoid the explanatory problem of the thing interpretation of nuclear models. Similarly, these models are each fine-tuned in different ways by the experimenter in order to reflect the specific experimental situation for which they are being used. The liquid-drop model will never be fine-tuned to better reflect the spectral emissions of excited nucleons because it was never intended to explain these processes. Again, nowhere does the liquid-drop model deny or contradict the processes that are used in the shell model to model spectral emission and beta decay. In short, nowhere do the ontologies of these two models differ. Rather, it is because this single, processual ontology is used for two different scientific explanations that we have two different models.

5.4 Conclusion and Prospectus

We have now seen, once more, that an apparent thing can be redescribed and explained in terms of processes. In the case of the nucleus, as opposed to the cases of the molecule and the candle flame, we have also discovered that the supposed thing underliers of these defining and explaining processes are only things if one stretches the definition. Protons, neutrons, and electrons are all dubiously things at all. They are non-local and non-localizable, and they are all defined as fluctuations in an infinite field. We therefore conclude that the nucleus is not a thing, but a collection of processes. We also suspect that this collection of processes will have no thing underliers.

We saw a further benefit in the process-realist account of the nucleus. That is, the explanatory processes of the nuclear system are robust and non-contradictory across nuclear models. This is in stark contrast to the thing interpretation of the nuclear models, which produced irredeemable contradictions between the models. The thing interpretation of the liquid-drop and shell models produced two different nuclei with different and incompatible structures, properties, and even haeccity. In providing a non-contradictory interpretation of these models, the process realist once again goes beyond mere parity with thing realism. I.e., process realism is once again shown to be superior in its account of scientific models.

Finally, we must note that the key difference between process and thing ontologies, in the context of models of the nucleus at least, is that they characterize models as explaining fundamentally different sorts of features of the nuclear system. The thing interpretation of the liquid-drop and shell models characterized these models as offering explanations of static properties, structures, and thing-components: the shape of the nucleus, the magic numbers, etc. This is evidently not the purpose of these models, both historically and in contemporary uses. The models are used to explain behaviors (i.e., processes) first. Insofar as the nuclear system has certain behaviors that we *describe* in terms of these static properties—e.g., the way we use the shape of the nucleus to characterize the stages of oscillation within the processes of fission—these static properties are still useful tools. However, they are not explanatory and must not be reified. Only the behaviors of the system are of interest to those that use the model.

The process interpretation reorients our analysis to describe these behaviors. We interpret the liquid-drop and shell models as tools to explain fission and excitation/decay processes, just as they were originally intended historically. In so doing, the process interpretation also reorients our interpretation of these models to match the experimental use to which they are put. These models are not meant to be descriptions of the static being of the world, but rather are constructed to describe, at their most basic level, specific experiments. The liquid-drop model is meant to describe fission, neutron scattering, and neutron capture in neutron bombardment experiments. The shell model is meant to describe nucleon excitation and decay resulting in spectral emission and nuclear decay processes in decay and spectral line experiments. Experiments are dynamic first and foremost. The experimenter acts on the system, and observes some change in the system that results from the dynamic sequence their intervention triggers. It is only natural, then, to expect that models meant to describe these experiments are about those system dynamics, those processes. Inference to anything else is unwarranted.

Summary and Prospectus

While process realism as I have called it is strongly supported by the necessary existence of processes to ontologically ground our experiences (observations, experiments, etc.), its true strength comes from the means by which it balances interpretive strength, historical and practical accuracy, and epistemic modesty. In making use of process realism, we can resolve many of the problems in the interpretation of scientific theories. We also match more closely the language used by working physicists throughout history, and make evident the core practical challenges of physical-theory construction. Finally, we gain all of this with far less epistemic risk than thing interpretations incur, being conventionally known as resting on a miracle. The process realist needs only to notice that we cannot physically observe or experiment at all if there are no physical processes for us to engage in.

However, while I have suggested throughout that process realism is superior as an interpretive metaphysics for the practice and theory of science in general, there is still much work to be done. In particular, we notice three key areas for enhancing our understanding of science in tandem with improvements to process realism.

First, while I have suggested here that process realism can co-opt the thing realist's robustness arguments, a full analysis of robustness within process-ontological interpretations of science should be attempted. This will trade on the addition of nuance to the idea of relative levels of stability I offered in Chapter 2.

Second, process realism remains one of the least studied ontologies historically. This is especially evident when we consider that many of the historical proponents of process ontology, or proponents of process-adjacent philosophy, appear in cultures outside of historical Europe. I have collected some isolated clusters here to mark points where these works could be integrated, but a thorough analysis is needed to fully grasp the connections between the development of science in Europe, the rejection of orthodox Christian (substance) ontologies, and the influence of Chinese and Japanese philosophies in this developmental process. The evidence of some connection is overwhelming, given the key importance of, e. g., Yukawa and Tomanaga on the development of quantum field theory and Needham's remarks on the social impacts of Chinese engineering and medicine on European thought in the early modern period.

Finally, process realism faces a serious challenge in the construction of spatial notions. Ideas of locality and non-locality in particular are somewhat troubling for the process realist, but remain integral to much scientific work. Thus, just like the problem of time and change for substance/thing realism, process realism faces a problem of space. I have suggested here that stability can and should be understood purely as a relation between processes. This suggests a way forward for the process realist in constructing spatial notions. However, the full account must be left for another work.

Bibliography

Aristotle (1984). *The Complete Works of Aristotle. The Revised Oxford Translation*, vols. 1 and 2. Ed. J. Barnes. Princeton, NJ: Princeton University Press.
Aristotle, *Categories*. Translated in *The Complete Works of Aristotle. The Revised Oxford Translation*. Ed. J. Barnes. Princeton, NJ: Princeton University Press (1984). Vol 1: 3–24.
Aristotle, *Metaphysics*. Translated in *The Complete Works of Aristotle. The Revised Oxford Translation*. Ed. J. Barnes. Princeton, NJ: Princeton University Press (1984). Vol 2: 1552–1728.
Aristotle, *Physics*. Translated in *The Complete Works of Aristotle. The Revised Oxford Translation*. Ed. J. Barnes. Princeton, NJ: Princeton University Press (1984). Vol 1: 315–446.
Atkinson, David (2006). "A Relativistic Zeno Effect." *Synthese* 160(1): 5–12.
Ayers, Michael (1990). *Locke*, vol. 2. London: Routledge.
Baker, David J. (2009). "Against Field Interpretations of Quantum Field Theory." *British Journal for the Philosophy of Science*, 60(3): 585–609.
Baker, David John (2010). "Symmetry and the Metaphysics of Physics." *Philosophy Compass* 5(12): 1157–1166.
Barnes, Elizabeth C. (2002). "The Miraculous Choice Argument for Realism." *Philosophical Studies*, 111(2): 97–120. DOI: 10.1023/A:1021204812809.
Barwich, Ann-Sophie (2018). "Measuring the World: Olfaction as a Process Model of Perception." In D. Nicholson and J. Dupré (eds.) *Everything Flows: Towards a Processual Philosophy of Biology*. Oxford: Oxford University Press. 337–356.
Becher, Johann J. (1669). *Physica Subterranea*. Hachette Livre-BNF: 1733rd edition (February 21, 2022).
Becher, Johann J. (1708). *True Theory of Medicine*.
Belot, Gordon (2015). "Down to Earth Underdetermination." *Philosophy and Phenomenological Research*, 91: 455–464.
Bennett, Jonathan (1965). "Substance, Reality, and Primary Qualities." *American Philosophical Quarterly*, 2(1): 1–17.
Bennett, Karen (2017). *Making Things Up*. New York: Oxford University Press.
Bergson, Henri (1994 [1896]). *Matter and Memory*. Trans. N.M. Paul and W.S. Palmer. New York: Zone Books.
Bickhard, Mark (2009). "Interactivism: A Manifesto." *New Ideas in Psychology*, 27: 85–95.
Bogen, Jim and Jim Woodward (1988). "Saving the Phenomena." *Philosophical Review*, 97(3): 303–352.
Bohr, Aage and Ben Mottelson (1969). *Nuclear Structure*. New York: Benjamin.
Bohr, Niels (1936). "Neutron Capture and Nuclear Constitution." *Nature*, 137: 344–348.
Bohr, Niels (1937). "Transmutations of Atomic Nuclei." *Science*, 86: 161–165.
Bokulich, Alisa (forthcoming). "Towards a Taxonomy of Model-Ladenness of Data." *PSA: Proceedings of the Biennial Meeting of the Philosophy of Science Association*.
Bokulich, Alisa and Wendy Parker (2021). "Data Models, Representation, and Adequacy-for-Purpose." *European Journal for Philosophy of Science*, 11(1): 31–64.
Boniolo, Giovanni, Carlo Petrovich, and Gualtiero Pisent (2002). "Notes on the Philosophical Status of Nuclear Physics." *Foundations of Science*, 7(4): 425–452.
Botterell, Andrew (2004). "Temporal Parts and Temporary Intrinsics." *Metaphysica*, 5(2): 5–23.
Boyd, Richard N. (1989). "What Realism Implies and What It Does Not." *Dialectica*, 43(1–2): 5–29. DOI: 10.1111/j.1746–8361.1989.tb00928.

Brading, Katherine and Elaine Landry (2006). "Scientific Structuralism: Presentation and Representation." *Philosophy of Science*, 73: 571–581.

Broad, Charlie Dunbar (1959 [1934]). "Autobiography." In P.A. Schilpp (ed.) *The Philosophy of C.D. Broad*. New York: Tudor. 3–68.

Brower, Jeffrey E. (2010). "Aristotelian Endurantism: A New Solution to the Problem of Temporary Intrinsics." *Mind*, 119: 883–905.

Brown, James R. (1982). "The Miracle of Science." *Philosophical Quarterly*, 32(128): 232–244. DOI: 10.2307/2219325.

Brown, Robert (1828). "A Brief Account of Microscopical Observations Made in the Months of June, July, and August 1827, on the Particles Contained in the Pollen of Plants; and on the General Existence of Active Molecules in Organic and Inorganic Bodies." *Edinburgh New Philosophical Journal*, 5: 358–371.

Brush, Stephen G. (1968). "A History of Random Processes: I. From Brown to Perrin." *Archive for History of Exact Sciences*, 5(1): 1–36.

Bueno, Otávio (1999). "What Is Structural Empiricism? Scientific Change in an Empiricist Setting." *Erkenntnis*, 50: 59–85.

Bueno, Otávio (2000). "Empiricism, Scientific Change and Mathematical Change." *Studies in History and Philosophy of Science Part A*, 31(2): 269–296.

Bueno, Otávio (2001). "Weyl and von Neumann: Symmetry, Group Theory, and Quantum Mechanics." Available at Philsci Archive. Philsci-archive.pitt.edu/409/ (last accessed June 1, 2020).

Bueno, Otávio (2006). "The Methodological Character of Symmetry Principles." *Abstracta*, 3: 3–28.

Bueno, Otávio (2016). "Belief Systems and Partial Spaces." *Foundations of Science*, 21: 225–236.

Bueno, Otávio (2019). "Can Quantum Objects Be Tracked?" In O. Bueno, R-L Chen, and M. Fagan (eds.) *Individuation, Process, and Scientific Practices*. Oxford: Oxford University Press, pp. 239–258.

Bueno, Otávio, Ruey-Lin Chen, and Melinda B. Fagan (2019). *Individuation, Process, and Scientific Practices*. Oxford: Oxford University Press.

Bueno, Otávio and Steven French (2011). "How Theories Represent." *British Journal for the Philosophy of Science*, 62(4): 857–894.

Bueno, Otávio and Steven French (2012). "Can Mathematics Explain Physical Phenomena?" *British Journal for the Philosophy of Science*, 63: 85–113.

Bueno, Otávio and Steven French (2020). *Applying Mathematics: Immersion, Inference, and Interpretation*. Oxford: Oxford University Press.

Bueno, Otávio, Steven French, and James Ladyman (2002). "On Representing the Relationship between the Mathematical and the Empirical." *Philosophy of Science*, 69: 497–518.

Busch, Jacob (2008). "No New Miracles, Same Old Tricks." *Theoria*, 74(2): 102–114. DOI: 10.1111/j.1755–2567.2008.00011.x.

Butterfield, Jeremy (2005). "On Symmetry and Conserved Quantities in Classical Mechanics." In: Demopoulos, W., Pitowsky, I. (eds) *PhysicalTheory and its Interpretation*. The Western Ontario Series in Philosophy of Science, Vol 72. Springer, Dordrecht. https://doi.org/10.1007/1-4020-4876-9-3, pp. 43–100.

Carnap, Rudolph (1934). *Logische Syntax der Sprache*. Schriften zur wissenschaftlichen Weltauffassung 8. Vienna: Julius Springer.

Carnap, Rudolph (1987). "On Protocol Sentences." Trans. Richard Creath and Richard Nollan. *Noûs*, 21(4): 457–470.

Carter, W.R. (1989). "How to Change Your Mind." *Canadian Journal of Philosophy*, 19: 1–14.

Cartwright, Nancy (1983). *How the Laws of Physics Lie*. Oxford: Clarendon Press.
Cartwright, Nancy (1991). "Replicability, Reproducibility, and Robustness: Comments on Harry Collins." *History of Political Economy*, 23(1): 143–155.
Cartwright, Nancy (2001). "What Is Wrong with Bayes Nets?" *The Monist*, 84(2): 242–264.
Cartwright, Nancy (2002). "Against Modularity, the Causal Markov Condition and Any Link Between the Two." *British Journal of Philosophy of Science*, 53: 411–453.
Cartwright, Nancy (2006). "From Metaphysics to Method: Comments on Manipulability and the Causal Markov Condition." *British Journal of Philosophy of Science*, 57: 197–218.
Caurier, Etienne, Gabriel Martínez-Pinedo, Frederic Nowacki, Alfredo Poves, and Andres Zuker (2005). "The Shell Model as a Unified View of Nuclear Structure." *Reviews of Modern Physics*, 77: 427–488.
Cavendish, Henry (2011 [1766]). *The Scientific Papers of the Honourable Henry Cavendish, F.R.S* (Cambridge Library Collection – Physical Sciences) Ed. J. Maxwell and S. Larmor. Cambridge: Cambridge University Press. DOI: 10.1017/CBO9780511722417.
Chakravartty, Anjan (2007). *A Metaphysics for Scientific Realism: Knowing the Unobservable*. Cambridge: Cambridge University Press.
Chakravartty, Anjan (2008). "What You Don't Know Can't Hurt You: Realism and the Unconceived." *Philosophical Studies*, 137: 149–158.
Chalmers, Alan (2009). *The Scientist's Atom and the Philosopher's Stone: How Science Succeeded and Philosophy Failed to Gain Knowledge of Atoms*. (Boston Studies in the Philosophy of Science 279). Dordrecht: Springer.
Chalmers, Alan (2011). "Drawing Philosophical Lessons from Perrin's Experiments on Brownian Motion: A Response to van Fraassen." *British Journal for the Philosophy of Science*, 62(4): 711–732.
Chaudesaigues (1908). "Le Mouvement Brownien et le Formule d'Einstein." *Comptes Rendus*, 147: 1044–1046.
Chen, Ruey-Lin (2018). "Experimental Individuation: Creation and Presentation." In Otávio Bueno, Ruey-Lin Chen, and Melinda Bonnie Fagan (eds.) *Individuation, Processes, and Scientific Practices*. New York: Oxford University Press, pp. 192–213.
Chirimuuta, M. (2015). *Outside Colour: Perceptual Science and the Puzzle of Colour in Philosophy*. Cambridge, MA: MIT Press.
Clausius, Rudolf (1851). "On the Moving Force of Heat, and the Laws Regarding the Nature of Heat Itself Which Are Deducible Therefrom." *Philosophical Magazine*, 2(4): 1–21, 102–119.
Clayton, Philip (2008). "Introduction to Process Thought." In T. Eastman and H. Keeton (eds.) *Physics and Whitehead: Quantum Process and Experience*. Albany, NY: State University of New York Press, pp. 3–13.
Cohen, Sheldon (1984). "Aristotle's Doctrine of the Material Substrate." *Philosophical Review*, 93(2): 171–194.
Coko, Klodian (2019). "Towards a Mutually Beneficial Integration of History and Philosophy of Science: The Case of Jean Perrin." In E. Herring, K.M. Jones, K.S. Kiprijanov, and L.M. Sellers (eds.) *The Past, Present, and Future of Integrated History and Philosophy of Science*. London: Routledge, pp. 186–209.
Coko, Klodian (2020). "Jean Perrin and the Philosophers' Stories: The Role of Multiple Determination in Determining Avogadro's Number." *Hopos: The Journal of the International Society for the History of Philosophy of Science*, 10(1): 143–193.

Cook, Norman D. (2006). *Models of the Atomic Nucleus: Unification through a Lattice of Nucleons.* Berlin: Springer.

Crupi, Vincenzo, Roberto Festa, and Tommaso Mastropasqua (2008). "Bayesian Confirmation by Uncertain Evidence: A Reply to Huber [2005]: Articles." *British Journal for the Philosophy of Science*, 59(2): 201–211.

Crupi, Vincenzo, Branden Fitelson, and Katya Tentori (2007). "Probability, Confirmation, and the Conjunction Fallacy." *Thinking and Reasoning*, 14(2): 182–199.

Da Costa, Newton C.A. and Steven French (1990). "The Model-Theoretic Approach to the Philosophy of Science." *Philosophy of Science*, 57(2): 248–265. DOI: 10.1086/289546.

Da Costa, Newton C.A. and Steven French (2003). *Science and Partial Truth: A Unitary Approach to Models and Scientific Reasoning.* Oxford: Oxford University Press.

Davidson, Donald (2001). "The Very Idea of a Conceptual Scheme." In Donald Davidson, *Inquiries into Truth and Interpretation.* Oxford: Clarendon, pp. 183–198.

DeGrazia, David (2005). *Human Identity and Bioethics.* New York: Cambridge University Press.

Dellsén, Finnur (2016). "Explanatory Rivals and the Ultimate Argument." *Theoria*, 82(3): 217–237. DOI: 10.1111/theo.12084.

DeVries, Willem A. (2005). *Wilfrid Sellars.* Chesham, Bucks: Acumen/Montreal and Kingston: McGill-Queen's University Press.

Dirac, Paul (1955). Address to the Indian Science Congress in Baroda.

Dowe, Phil (1992). "Wesley Salmon's Process Theory of Causation and the Conserved Quantity Theory." *Philosophy of Science*, 59: 195–216.

Dowe, Phil (1995). "Causality and Conserved Quantities: A Reply to Salmon." *Philosophy of Science*, 62(2): 321–333.

Dowe, Phil (2000). *Physical Causation.* Cambridge: Cambridge University Press.

Dowe, Phil (2003). "Physical Causation." *Philosophy and Phenomenological Research*, 67(1): 244–248.

Duhem, Pierre (1954 [1914]). *The Aim and Structure of Physical Theory.* Trans. from 2nd edition by P.W. Wiener; originally published as *La Théorie Physique: Son Objet et sa Structure* (Paris: Marcel Riviera & Cie.). Princeton, NJ: Princeton University Press.

Dupré, John (2010). *Processes of Life: Essays in the Philosophy of Biology.* Oxford: Oxford University Press.

Dupré, John (2014). *Processes of Life: Essays in the Philosophy of Biology.* Oxford: Oxford University Press.

Dupré, John (2018). "Processes, Organisms, Kinds, and the Inevitability of Pluralism." In Otávio Bueno, Ruey-Lin Chen, and Melinda Bonnie Fagan (eds.) *Individuation, Processes, and Scientific Practices.* New York: Oxford University Press, pp. 21–38.

Earley, Joseph (2008a). "Process Structural Realism, Instance Ontology and Societal Order." In Franz Riffert and Hans-Joachim Sander (eds.) *Researching with Whitehead: System and Adventure.* Berlin: Alber, pp. 190–211.

Earley, Joseph (2008b). "Constraints on the Origin of Coherence in Far-From-Equilibrium Chemical Systems." In T. Eastman and H. Keeton (eds.) *Physics and Whitehead: Quantum Process and Experience.* Albany, NY: State University of New York Press, pp. 63–73.

Earley, Joseph (2008c). "Ontologically Significant Aggregation: Process Structural Realism (PSR)." In M. Weber (ed.) *Handbook of Whiteheadian Process Thought.* Berlin and New York: De Gruyter, pp. 2–179.

Earley, Joseph (2008d). "How Philosophy of Mind Needs Philosophy of Chemistry." Preprint available on Philsci Archive. URL: http://philsci-archive.pitt.edu/4031/ (last accessed 18 January 2023).

Earley, Joseph (2012). "A Neglected Aspect of the Puzzle of Chemical Structure: How History Helps." *Foundations of Chemistry*, 14(3): 235–243.

Earley, Joseph (2016). "How Properties Hold Together in Substances." In Eric Scerri and Grant Fisher (eds.) *Essays in Philosophy of Chemistry.* New York: Oxford University Press, pp. 199–216.

Earman, John (2002). "Laws, Symmetry, and Symmetry Breaking; Invariance, Conservation Principles, and Objectivity." *Proceedings of the 2002 Biennial Meeting of the Philosophy of Science Association. Part II: Symposia Papers* (Ed. Sandra D. Mitchell) 71(5): 1227–1241.

Eastman, Tim and Hank Keeton (2008). *Physics and Whitehead: Quantum Process and Experience.* Albany, NY: State University of New York Press.

Eberhardt, Frederick and Richard Scheines (2006). "Interventions and Causal Inference." *Conference Proceedings of the PSA 2006.*

Ehring, Douglas (1997). "Lewis, Temporary Intrinsics and Momentary Tropes." *Analysis*, 57: 254–258.

Ehring, Douglas (2001). "Temporal Parts and Bundle Theory." *Philosophical Studies*, 104: 163–168.

Einheuser, Iris (2012). "Is There a (Meta)Problem of Change?" *Analytic Philosophy*, 53(4): 344–351.

Einstein, Albert (1905a). "On a Heuristic Viewpoint Concerning the Production and Transformation of Light." *Annalen der Physik*, 17: 132–148.

Einstein, Albert (1905b). "A New Determination of Molecular Dimensions." *Annalen der Physik*, 19: 289–305.

Einstein, Albert (1905c). "On the Motion of Small Particles Suspended in Liquids at Rest Required by the Molecular-Kinetic Theory of Heat." *Annalen der Physik*, 17: 549–560.

Einstein, Albert (1905d). "On the Electrodynamics of Moving Bodies." *Annalen der Physik*, 17: 891–921.

Einstein, Albert (1905e). "Does the Inertia of a Body Depend Upon its Energy Content?" *Annalen der Physik*, 18: 639–641.

Eronen, Markus (2015). "Robustness and Reality." *Synthese*, 192(12): 3961–3977.

Faraday, Michael (2016 [1848]). "The Chemical History of a Candle." Series of lectures reprinted in B. Hammack and D. DeCoste (eds.) *Michael Faraday's The Chemical History of a Candle with Guides to Lectures, Teaching Guides, and Student Activities.* Urbana: Articulate Noise Books

Ferner, Adam and Thomas Pradeu (2017). "Ontologies of Living Beings." *Philosophy, Theory, and Practice in Biology*, 9(4).

Feyerabend, Paul K. (1962). "Explanation, Reduction, and Empiricism." In Herbert Feigl and Grover Maxwell (eds.) *Scientific Explanation, Space, and Time.* Minneapolis: University of Minnesota Press, pp. 28–97.

Feyerabend, Paul K. (1985 [1969]). "Science Without Experience." In P.K. Feyerabend, *Realism, Rationalism, and Scientific Method* (Philosophical Papers 1). Cambridge: Cambridge University Press, pp. 132–136.

Fine, Kit (1994). "Essence and Modality: The Second Philosophical Perspectives Lecture." *Philosophical Perspectives*, 8: 1–16.

Finkelstein, David (1996). *Quantum Relativity: A Synthesis of the Ideas of Einstein and Heisenberg.* Berlin: Springer.

Finkelstein, David (2008). "Physical Process and Physical Law." In T. Eastman and H. Keeton (eds.) *Physics and Whitehead: Quantum Process and Experience.* Albany, NY: State University of New York Press, pp. 180–186.

Fiocco, M. Oreste (2010). "Temporary Intrinsics and Relativization." *Pacific Philosophical Quarterly*, 91(1): 64–77.

Fraser, Doreen (2011). "How to Take Particle Physics Seriously: A Further Defence of Axiomatic Quantum Field Theory." *Studies in History and Philosophy of Modern Physics*, 42: 126–135.

French, Steven (2000). "The Reasonable Effectiveness of Mathematics: Partial Structures and the Application of Group Theory to Physics." *Synthese*, 125: 103–120.

French, Steven (2001). "Symmetry, Structure and the Constitution of Objects." *Conference on Symmetries in Physics: New Reflections*, University of Oxford, January 2001, Available at Philsci Archive, philsci-archive-dev.library.pitt.edu/327/ (last accessed February 13, 2023).

French, Steven (2003a). "A Model-Theoretic Account of Representation." *Philosophy of Science*, 70(5): 1472–1483.

French, Steven (2003b). "Scribbling on the Blank Sheet: Eddington's Structuralist Conception of Objects." *Studies in History and Philosophy of Modern Physics*, 34: 227–259.

French, Steven (2006). "Structure as a Weapon of the Realist." *Proceedings of the Aristotelian Society*, 106: 1–19.

French, Steven (2010). "The Interdependence of Structure, Objects and Dependence." *Synthese*, 175 (S1): 89–109.

French, Steven (2011). "Shifting to Structures in Physics and Biology: A Prophylactic for Promiscuous Realism." *Studies in History and Philosophy of Science, Part C: Studies in History and Philosophy of Biological and Biomedical Sciences*, 42(2): 164–173.

French, Steven (2016). "Identity Conditions, Idealisations and Isomorphisms: A Defence of the Semantic Approach." *Synthese*, 198(Sup 24): 1–21.

Frigg, Roman (2006). "Scientific Representation and the Semantic View of Theories." *Theoria*, 55: 37–53.

Frigg, Roman (2010). "Fiction and Scientific Representation." In R. Frigg and M. Hunter (eds.) *Beyond Mimesis and Nominalism: Representation in Art and Science*. Springer, pp. 97–138.

Frisch, Otto R. (1934). "Induced Radioactivity of Sodium and Phosphorus." *Nature*, 133: 721–722.

Frost-Arnold, Greg (2010). "The No-Miracles Argument for Realism: Inference to an Unacceptable Explanation." *Philosophy of Science*, 77(1): 35–58. DOI: 10.1086/650207.

Gamow, George (1929). "Über die Structur des Atomkernes." *Physikalische Zeitschrift*, 30: 717–720.

Garrett, Brian (1998). *Personal Identity and Self-Consciousness*, London: Routledge.

Gill, Mary Louise (1989). *Aristotle on Substance: The Paradox of Unity*. Princeton, NJ: Princeton University Press.

Glymour, Clark (1975). "Relevant Evidence." *The Journal of Philosophy*, 72(14): 403–426.

Glymour, Clark (1980). *Theory and Evidence*. Princeton, NJ: Princeton University Press.

Glymour, Clark (2013). "Theoretical Equivalence and the Semantic View of Theories." *Philosophy of Science*, 80, pp. 286–297.

Gómez, Omar S., Natalia Juristo, and Sira Vegas (2010). "Replications Types in Experimental Disciplines." In *Proceedings of the 2010 ACM-IEEE International Symposium on Empirical Software Engineering and Measurement – ESEM '10*. Bolzano: ACM Press. DOI: 10.1145/1852786.1852790.

Gooday, Graeme and Donald Mitchell (2013). "Rethinking Classical Physics." In J.Z. Buchwald and R. Fox (eds.) *The Oxford Handbook of the History of Physics*. Oxford: Oxford University Press, pp. 721–764.

Goodman, Nelson (1955). *Fact, Fiction, and Forecast*, Indianapolis, IN: Bobbs-Merrill.

Goodman, Nelson (1977). *The Structure of Appearance*, 3rd edition. Dordrecht: Reidel.

Gorham, Geoffrey (2010). "Descartes on Persistence and Temporal Parts." In J.K. Campbell, M. O'Rourke, and H. Silverstein (eds.) *Time and Identity*. Cambridge, MA: MIT Press, pp. 165–182.

Griesemer, James (2018). "Individuation of Developmental Systems: A Reproducer Perspective." In Otávio Bueno, Ruey-Lin Chen, and Melinda Bonnie Fagan (eds.) *Individuation, Processes, and Scientific Practices*. New York: Oxford University Press, pp. 137–164.

Guay, Alexander and Thomas Pradeu (forthcoming). "Right Out of the Box: How to Situate Metaphysics of Science in Relation to Other Metaphysical Approaches." *Synthese* special issue: "New Metaphysics of Science," ed. Max Kistler.

Guay, Alexander and Olivier Sartenaer (2018). "Emergent Quasiparticles: Or, How to Get a Rich Physics from a Sober Metaphysics." In Otávio Bueno, Ruey-Lin Chen, and Melinda Bonnie Fagan (eds.) *Individuation, Processes, and Scientific Practices*. New York: Oxford University Pres, pp. 214–238.

Guay, Alexandre and Thomas Pradeu (2015). "To Be Continued: The Genidentity of Physical and Biological Processes." In Alexandre Guay and Thomas Pradeu (eds.) *Individuals Across the Sciences*. New York: Oxford University Press, pp. 317–347.

Hacking, Ian (1984). "Representing and Intervening." *British Journal for the Philosophy of Science*, 35(4): 381–390.

Halvorson, Hans (2012). "What Scientific Theories Could Not Be." *Philosophy of Science*, 79: 183–206.

Hansson, Tobias (2007). "The Problem(s) of Change Revisited." *Dialectica*, 61(2): 265–274.

Hartmann, Stephan (2005). "The World as a Process: Simulations in the Natural and Social Sciences." In R. Hegselmann et.al. (eds.) *Simulation and Modelling in the Social Sciences from the Philosophy of Science Point of View*. Dordrecht: Kluwer, pp. 77–100.

Haslanger, Sally (1989a). "Persistence, Change and Explanation." *Philosophical Studies*, 56: 1–28.

Haslanger, Sally (1989b). "Endurance and Temporary Intrinsics." *Analysis*, 49: 119–125.

Hausman, Daniel and Jim Woodward (1999). "Independence, Invariance, and the Causal Markov Condition." *British Journal for the Philosophy of Science*, 50: 521–583.

Hausman, Daniel and Jim Woodward (2004). "Modularity and the Causal Markov Condition: A Restatement." *British Journal for the Philosophy of Science*, 55: 147–161.

Hawley, Katherine (2001). *How Things Persist*. Oxford: Oxford University Press.

Heisenberg, Werner (1932a). "Über den Bau der Atomkerne I." *Zeitschrift für Physik*, 77: 1–11.

Heisenberg, Werner (1932b). "Über den Bau der Atomkerne II." *Zeitschrift für Physik*, 78: 156–164.

Heisenberg, Werner (1932c). "Über den Bau der Atomkerne III." *Zeitschrift für Physik*, 80: 587–596.

Heller, Mark (1992). "Things Change." *Philosophy and Phenomenological Research*, 52: 695–704.

Hempel, Carl G. (1952). "Fundamentals of Concept Formation in Empirical Science." Reprinted in O. Neurath, R. Carnap, and C. Morris (eds.) *Foundations of the Unity of Science*, vol. 2. Chicago: University of Chicago Press 1970, pp. 651–746.

Hendrick, Clyde (1991). "Replication, Strict Replications, and Conceptual Replications: Are They Important?" *Journal of Social Behavior and Personality*, 5(4):41–49.

Hinchliff, Mark (1996). "The Puzzle of Change." *Philosophical Perspectives*, 10: 119–136.

Hippocrates (2013). "On Airs, Waters and Places" In Evan Hayes and Stephen Nimis (eds.) *Hoppocrates' On Air's Waters, and Places and the Hippocratic Oath: An Intermediate Greek Reader*. Faenum Publishing Ltd.

Hitchcock, Christopher (2001a). "The Intransitivity of Causation Revealed in Equations and Graphs." *The Journal of Philosophy*, 98(6): 273–299. DOI: 10.2307/2678432.

Hitchcock, Christopher (2001b). "A Tale of Two Effects" *Philosophical Review*, 110(3): 361–396. DOI: 10.1215/00318108-110-3-361.

Hitchcock, Christopher (2006). "On the Importance of Causal Taxonomy." In A. Gopnik and L. Schulz (eds.) *Causal Learning: Psychology, Philosophy and Computation.* New York: Oxford University Press, pp. 101–114.

Hitchcock, Christopher (2007a). "Prevention, Preemption, and the Principle of Sufficient Reason." *Philosophical Review*, 116(4): 495–532. DOI: 10.1215/00318108-2007-012.

Hitchcock, Christopher (2007b). "What Russell Got Right." In Huw Price ad Richard Corry (eds.), *Causation, Physics, and the constitution of Reality: Russell's Republic Revisited.* Oxford, Oxford University Press, pp. 45–65.

Hitchcock, Christopher and Jim Woodward (2003). "Explanatory Generalizations, Part II: Plumbing Explanatory Depth." *Nôus*, 37(2): 181–199. DOI: 10.1111/1468-0068.00435.

Hofweber, Thomas (2009). "The Meta-Problem of Change." *Noûs*, 43(2): 286–314.

Howson, Colin (2000). *Hume's Problem: Induction and the Justification of Belief.* Oxford: Oxford University Press. DOI: 10.1093/0198250371.001.0001.

Hudson, Hud (2001). *A Materialist Metaphysics of the Human Person*, Ithaca: Cornell University Press.

Hudson, Hud (2007). "I Am Not an Animal!" In P. van Inwagen and D. Zimmerman (eds.) *Persons: Human and Divine.* Oxford: Oxford University Press, pp. 216–234.

Hudson, Robert G. (2014). *Seeing Things: The Philosophy of Reliable Observation.* Oxford: Oxford University Press.

Hudson, Robert G. (2020). "The Reality of Jean Perrin's Atoms and Molecules." *British Journal for the Philosophy of Science*, 71(1): 33–58.

Ismael, Jenann and Jonathan Schaffer (2020). "Quantum Holism: Nonseparability as Common Ground." *Synthese*, 197(10): 4131–4160.

James, William (1977 [1909]). *A Pluralistic Universe.* Cambridge, MA: Harvard University Press.

James, William (1981 [1890]). *The Principles of Psychology.* Cambridge, MA: Harvard University Press.

Janssen, Michel (2004a). "Relativity." In M C. Horowitz et. al. (eds.) *New Dictionary of the History of Ideas.* New York: Scribner's, pp. 2039–2047.

Janssen, Michel (2004b) "L'Ottica e l'elettrodinamica dei corpi in movimento." With John Stachel. In Sandro Petruccioli et al. (eds.) *Storia Della Scienza*, vol. 8. Rome: Istituto della Enciclopedia Italiana, pp. 363–379.

Janssen, Michel (2011). "Lorentz als wegbereider voor de speciale relativiteitstheorie." With Anne J. Kox. *Nederlands Tijdschrift voor Natuurkunde*, 77: 344–347.

Jenkins, C.S. (2013). "Explanation and Fundamentality." In *Hoeltje, Schnieder, & Steinberg (eds) Varieties of Dependence: Ontological Dependence, Grounding, Supervenience, Response-Dependence.* Philosophia, pp. 211–242.

Johnson, William E. (1921). *Logic*, vol. 1. Cambridge: Cambridge University Press.

Johnston, Mark (1984). *Particulars and Persistence.* PH.D Dissertation. Princeton.

Johnston, Mark (1987a). "Human Beings." *Journal of Philosophy*, 84: 59–83.

Johnston, Mark (1987b). "Is There a Problem About Persistence?" *Aristotelian Society Supplementary Volume*, 61: 433–439.

Johnston, Mark (2016). "Remnant Persons: Animalism's Undoing." In S. Blatti and P. Snowdon (eds.) *Animalism: New Essays on Persons, Animals, and Identity.* Oxford University Press, pp. 89–127.

Jungerman, John A. (2008). "Evidence for Process in the Physical World." In T. Eastman and H. Keeton (eds.) *Physics and Whitehead: Quantum Process and Experience.* Albany, NY: State University of New York Press, pp. 47–56.

Kaiser, Marie (2018). "Individuating Part-Whole Relations in the Biological World." In Otávio Bueno, Ruey-Lin Chen, and Melinda Bonnie Fagan (eds.) *Individuation, Processes, and Scientific Practices*. New York: Oxford University Press, pp. 63–90.

Kenny, Anthony (1963). *Actions, Emotions, and Will*. New York: Humanities Press.

Keppel, Geoffrey (1982). *Design and Analysis. A Researcher's Handbook*, 2nd edition. Englewood Cliffs, NJ: Prentice Hall.

Kirk, Geoffrey, John-Earle Raven, and Malcolm Schofield (1957). *The Presocratic Philosophers*. Cambridge: Cambridge University Press.

Kitcher, Philip (1981). "Explanatory Unification." *Philosophy of Science*, 48(4): 507–531.

Klein, Charles J. (1999). "Change and Temporal Movement." *American Philosophical Quarterly*, 36: 225–239.

Kostas, Gavroglu and Ana Simões (2012). *Neither Physics Nor Chemistry: A History of Quantum Chemistry*. Cambridge, MA: MIT Press.

Kuby, Daniel (2018). "Carnap, Feyerabend, and the Pragmatic Theory of Observation." *HOPOS: The Journal of the International Society for the History and Philosophy of Science*, 8(2): 432–470 DOI: 10.1086/698695.

Kuriyama, Shigehisa (1999). *The Expressiveness of the Body and the Divergence of Greek and Chinese Medicine*. New York: Zone Books.

Laudan, Larry (1981). "A Confutation of Convergent Realism." *Philosophy of Science*, 48: 19–48.

Laudan, Larry (1990). "Demystifying Underdetermination." In C. Wade Savage (ed.) *Scientific Theories*. Minneapolis: University of Minnesota Press, pp. 267–297.

Lavoisier, Antione (1770–1790) Collected Papers available at https://royalsocietypublishing.org/doi/pdf/10.1098/rspa.1947.0050 (last accessed 18 January 2023).

LeBihan, Saozig (2012). "Defending the Semantic View: What It Takes." *European Journal for the Philosophy of Science*, 2: 249–274.

Lewis, David (1971). "Counterparts of Persons and Their Bodies." *Journal of Philosophy*, 68: 203–211.

Lewis, David (1976). "Survival and Identity." In Amelie Rorty (ed.) *The Identities of Persons*. Berkeley, CA: University of California Press, pp. 117–140. Reprinted with significant postscripts in *Lewis's Philosophical Papers*, vol. 1. Oxford: Oxford University Press, pp. 117–140.

Lewis, David (1986). *On the Plurality of Worlds*. Oxford: Blackwell.

Lewis, Peter J. (2001). "Why the Pessimistic Induction Is a Fallacy." *Synthese*, 129(3): 371–380. DOI: 10.1023/A:1013139410613.

Lindberg, David C. (2007). *The Beginnings of Western Science: The European Scientific Tradition in Philosophical, Religious, and Institutional Context, Prehistory to A.D. 1450*, 2nd edition. Chicago: University of Chicago Press.

Lipton, Peter (1994). "Truth, Existence, and the Best Explanation." In A.A. Derksen (ed.) *The Scientific Realism of Rom Harré*. Tilburg: Tilburg University Press, pp. 89–111.

Lloyd, Elisabeth A. (2010). "Confirmation and Robustness of Climate Models." *Philosophy of Science*, 77(5): 971–984.

Lloyd, Elisabeth A. (2015). "Model Robustness as a Confirmatory Virtue: The Case of Climate Science." *Studies in History and Philosophy of Science Part A*, 49: 58–68.

Lombard, Lawrence B. (1994). "The Doctrine of Temporal Parts and the 'No-Change' Objection." *Philosophy and Phenomenological Research*, 54(2): 365–372.

Love, Alan C. (2018). "Individuation, Individuality, and Experimental Practice in Developmental Biology." In Otávio Bueno, Ruey-Lin Chen, and Melinda Bonnie Fagan (eds.) *Individuation, Processes, and Scientific Practices*. New York: Oxford University Press, pp. 165–191.

Lowe, Edward J. (1987). "Lewis on Perdurance versus Endurance." *Analysis*, 47: 152–154.
Lykken, David T. (1968). "Statistical Significance in Psychological Research." *Psychological Bulletin*, 70(3, Pt.1): 151–159. DOI: 10.1037/h0026141.
Lynds, Peter (2003). *Zeno's Paradoxes: A Timely Solution*. Available at PhilSci Archive, philsci-archive.pitt.edu/1197/ (last accessed August 1, 2022).
Lyon, Aiden and Mark Colyvan (2007). "The Explanatory Power of Phase Spaces." *Philosophia Mathematica*, 16(2), 227–243.
Lyons, Timothy D. (2003). "Explaining the Success of a Scientific Theory." *Philosophy of Science*, 70(5): 891–901. DOI: 10.1086/377375.
MacBride, Fraser (2001). "Four New Ways to Change Your Shape." *Australasian Journal of Philosophy*, 79: 81–89.
Mackie, David (1999). "Personal Identity and Dead People." *Philosophical Studies*, 95: 219–242.
Maël, Bathfield (2018). "Why Zeno's Paradoxes of Motion Are Actually about Immobility." *Foundations of Science*, 23(4): 649–679.
Magnus, P.D. and Craig Callender (2004). "Realist Ennui and the Base Rate Fallacy." *Philosophy of Science*, 71(3): 320–338. DOI: 10.1086/421536.
Malament, David (1996). "In Defense of Dogma: Why There Cannot Be a Relativistic Quantum Mechanical Theory of (Localizable) Particles." In R. Clifton (ed.) *Perspectives on Quantum Reality*. Dordrecht: Kluwer, pp. 1–10.
Malin, Shimon (2008). "Whitehead's Philosophy and the Collapse of Quantum States." In T. Eastman and H. Keeton (eds.) *Physics and Whitehead: Quantum Process and Experience*. Albany, NY: State University of New York Press, pp. 74–83.
Maxwell, James (1872). *Theory of Heat*. New York: D. Appleton & Co.
Mayer, Maria G. and J. Hans D. Jensen (1955). *Elementary Theory of Nuclear Shell Structure*. New York: Wiley.
Mayo, Deborah G. (1986). "Cartwright, Causality, and Coincidence." *PSA: Proceedings of the Biennial Meeting of the Philosophy of Science Association*, 1986: 42–58.
Mayo, Deborah G. (1996). *Error and the Growth of Experimental Knowledge*. Chicago: University of Chicago Press.
McCall, Storrs and Edward J. Lowe (2003). "3D/4D Equivalence, the Twins Paradox and Absolute Time." *Analysis*, 63: 114–123.
McKie, Douglas (1935). *Antoine Lavoisier: The Father of Modern Chemistry*. London: Victor Gollancz, ltd.
McTaggart, John Ellis (1908). "The Unreality of Time." *Mind*, 17(68): 457–474.
McTaggart, John Ellis (1927). *The Nature of Existence*, vol. 2. Cambridge: Cambridge University Press.
Meincke, Ann-Sophie (2018). "Autopoiesis, Biological Autonomy and the Process View of Life." *European Journal for the Philosophy of Science*, 9(1): 1–16.
Meincke, Ann-Sophie (2019). "The Disappearance of Change towards a Process Account of Persistence," *International Journal of Philosophical Studies*, 27(1): 12–30.
Meitner, Lise (1936). "Künstliche Umwandlungsprozesse beim Uran." In Egon Bretscher (ed.) *Kernphysik: Vorträge gehalten am Physikalischen Institut der Eidgenössischen Technischen Hochschule Zürich im Sommer 1936 (30. Juni – 4. Juli)*. Berlin: Springer.
Meitner, Lise and Otto R. Frisch (1939). "Disintegration of Uranium by Neutrons: A New Type of Nuclear Reaction." *Nature*, 143: 239–240.
Melia, Joseph (2000). "Continuants and Occurrents." *Proceedings of the Aristotelian Society*, Supplementary Volume 74: 77–92.
Mellor, David H. (1998). *Real Time II*. London: Routledge.

Menke, Cornelia (2014). "Does the Miracle Argument Embody a Base Rate Fallacy?" *Studies in History and Philosophy of Science*, 45: 103–108.

Merricks, Trenton (1994). "Endurance and Indiscernibility." *Journal of Philosophy*, 91: 165–184.

Mill, John S. (1900 [1867]). *A System of Logic, Ratiocinative and Inductive, Being a Connected View of the Principles of Evidence and the Methods of Scientific Investigation*. New York: Longmans, Green, and Co.

Moore, Alan W. (2012). *The Evolution of Modern Philosophy: The Evolution of Modern Metaphysics: Making Sense of Things*. Cambridge: Cambridge University Press.

Morganti, Matteo (2009). "Ontological Priority, Fundamentality and Monism." *Dialectica*, 63(3): 271–288. DOI: 10.1111/j.1746-8361.2009.01197.x.

Morrison, Margaret (2011). "One Phenomenon, Many Models: Inconsistency and Complementarity." *Studies in History and Philosophy of Science*, 42(2): 342–351.

Nagel, Thomas (1986). *The View from Nowhere*. New York: Oxford University Press.

Needham, Joseph (1969). *The Grand Titration*. New York: Routledge.

Ney, Alyssa (2015). "Fundamental Physical Ontologies and the Constraint of Empirical Coherence: A Defense of Wave Function Realism." *Synthese*, 192(10): 3105–3124. DOI: 10.1007/s11229-014-0633-9.

Noonan, Harold (2003). *Personal Identity*, 2nd edition. London: Routledge.

Noonan, Harold (2011). "The Complex and Simple Views of Personal Identity." *Analysis*, 71: 72–77.

North, Jill (2009). "The 'Structure' of Physics: A Case Study." *Journal of Philosophy*, 106: 57–88.

Norton, John (2003). "A Material Theory of Induction." *Philosophy of Science*, 70: 647–670.

Norton, John (2008). "Must Evidence Underdetermine Theory?" In M. Carrier, D. Howard, and J. Kourany (eds.) *The Challenge of the Social and the Pressure of Practice: Science and Values Revisited*. Pittsburgh: University of Pittsburgh Press, pp. 17–44.

Norton, John (2010). "There Are No Universal Rules for Induction." *Philosophy of Science*, 77: 765–777.

Norton, John (2015). "Replicability of Experiment." *Theoria*, 30(2): 229–248.

Norton, John (2021). *The Material Theory of Induction*. Complete manuscript available at http://www.pitt.edu/~jdnorton/homepage/research/ind_material.html (last accessed 18 January 2023).

Nosek, Brian A., Jeffrey R. Spies, and Matt Motyl (2012). "Scientific Utopia: II. Restructuring Incentives and Practices to Promote Truth Over Publishability." *Perspectives on Psychological Science*, 7(6): 615–631. DOI: 10.1177/1745691612459058.

Nye, Mary (1972). *Molecular Reality: A Perspective on the Scientific Work of Jean Perrin*. New York: American Elsevier.

Nye, Mary (ed.) (1984). *The Question of the Atom: From the Karlsruhe Congress to the First Solvay Conference. 1860–1911. A Selection of Primary Sources*. Los Angeles and San Francisco: Tomash.

Oderberg, David (2004). "Temporal Parts and the Possibility of Change." *Philosophy and Phenomenological Research*, 69(3): 686–703.

Olesko, Kathryn M. and Frederic L. Holmes (1994). "Experiment, Quantification and Discovery: Helmholtz's Early Physiological Researches, 1843–50." In D. Cahan (ed.) *Hermann Helmholtz and the Foundations of Nineteenth Century Science*. Berkeley: UC Press, pp. 50–108.

Olson, Eric (1997). *The Human Animal: Personal Identity without Psychology*. New York: Oxford University Press.

Olson, Eric and Karsten Witt (2019). "Narrative and Persistence." *Canadian Journal of Philosophy*, 49: 419–434.

Pais, Abraham (1991). *Niels Bohr's Times. In Physics, Philosophy, and Polity*. New York: Oxford University Press.

Papa-Grimaldi, Alba (1996). "Why Mathematical Solutions of Zeno's Paradoxes Miss the Point: Zeno's One and Many Relation and Parmenides' Prohibition." *The Review of Metaphysics*, 50 (December): 299–314.

Parfit, Derek (1971). "Personal Identity." *Philosophical Review*, 80: 3–27.

Parfit, Derek (1995). "The Unimportance of Identity." In H. Harris (ed.) *Identity*. Oxford: Oxford University Press, pp. 13–45.

Parfit, Derek (2012). "We Are Not Human Beings." *Philosophy*, 87: 5–28.

Parker, Wendy S. (2011). "When Climate Models Agree: The Significance of Robust Model Predictions." *Philosophy of Science*, 78(4): 579–600.

Paul, Laurie A. (2012). "Building the World from Its Fundamental Constituents." *Philosophical Studies*, 158(2): 221–256. DOI: 10.1007/s11098-012-9885-8.

Pemberton, John (2018). "Individuating Processes." In Otávio Bueno, Ruey-Lin Chen, and Melinda Bonnie Fagan (eds.) *Individuation, Processes, and Scientific Practices*. New York: Oxford University Press, pp. 39–62.

Peramatzis, Michail (2011). *Priority in Aristotle's Metaphysics*. Oxford University Press.

Perrin, Jean-Baptiste (1908). "L'agitation moléculaire et le movement Brownien." *Comptes rendus*, 146: 967–970.

Perrin, Jean-Baptiste (1910). *Brownian Movement and Molecular Reality*. Trans. Frederick Soddy. London: Taylor & Francis.

Perrin, Jean-Baptiste (1916). *Atoms*. Trans. Dalziel Hammick. 4th edition New York: Van Nostrand.

Perry, John (1972). "Can the Self Divide?" *Journal of Philosophy*, 69: 463–488.

Plato, *Cratylus*. In *Plato: Complete Works*. Ed. John Cooper. Indianapolis: Hackett (1997), pp. 101–156.

Plato, *Timaeus*. In *Plato: Complete Works*. Ed. John Cooper. Indianapolis: Hackett (1997), pp. 1224–1291.

Poincaré, Henri (1952 [1905]). *Science and Hypothesis*. New York: Dover.

Polanski, Marek (2009). "Goodman's Extensional Isomorphism and Syntactical Interpretations." *Theoria, An International Journal for Theory, History and Foundations of Science*, 24(2): 203–211.

Portides, Demetris (2011). "Seeking Representations of Phenomena: Phenomenological Models." *Studies in History and Philosophy of Science*, 42(2): 334–341.

Pradeu, Thomas (2018). "Genidentiy and Biological Processes." In D. Nicholson and J. Dupré (eds.) *Everything Flows: Towards a Processual Philosophy of Biology*. Oxford: Oxford University Press, pp. 96–112.

Pradeu, Thomas and Adam Ferner (2018). "Ontologies of Living Beings: Introduction." *Philosophy, Theory, and Practice in Biology* 9(4): 1–4.

Priestley, Joseph (1774). Collected Papers available on request at https://search.amphilsoc.org/collections/view?docId=ead/Mss.B.P931-ead.xml;query=;brand=default (last accessed 18 January 2023).

Psillos, Stathis (1994). "A Philosophical Study of the Transition from the Caloric Theory of Heat to Thermodynamics: Resisting the Pessimistic Meta-Induction." *Studies in the History and Philosophy of Science*, 25: 159–190.

Psillos, Stathis (1995). "Is Structural Realism the Best of Both Worlds?" *Dialectica*, 49: 15–46.

Psillos, Stathis (1996). "On van Fraassen's Critique of Abductive Reasoning." *Philosophical Quarterly*, 46(182): 31–47. DOI: 10.2307/2956303.

Psillos, Stathis (1999). *Scientific Realism: How Science Tracks Truth*. London: Routledge.

Psillos, Stathis (2001). "Is Structural Realism Possible?" *Philosophy of Science*, 68 (Supplementary Volume): S13–S24.

Psillos, Stathis (2011a). "Making Contact with Molecules: On Achinstein and Perrin." In G.J. Morgan (ed.) *Philosophy of Science Matters: The Philosophy of Peter Achinstein.* Oxford: Oxford University Press, pp. 177 – 190.
Psillos, Stathis (2011b). "Moving Molecules above the Scientific Horizon." *Journal for General Philosophy of Science*, 49: 339 – 363.
Putnam, Hilary (1975). *Mathematics, Matter and Method.* Cambridge: Cambridge University Press.
Putnam, Hilary (1978). *Meaning and the Moral Sciences.* London: Routledge.
Quine, Willard V.O. (1950). "Identity, Ostension and Hypostasis." In Willard V.O. Quine, *From a Logical Point of View.* Cambridge MA: Harvard University Press, pp. 65 – 79.
Quine, Willard V.O. (1960). *Word and Object.* Cambridge MA: MIT Press.
Quine, Willard V.O. (1970). *Philosophy of Logic.* New Jersey: Prentice Hall (2nd edition: 1986).
Quine, Willard V.O. (1975). "On Empirically Equivalent Systems of the World." *Erkenntnis*, 9: 313 – 328.
Quine, Willard V.O. (1990). "Three Indeterminacies." In R.B. Barrett and R.F. Gibson (eds.) *Perspectives on Quine.* Cambridge, MA: Blackwell, pp. 1 – 16.
Radder, Hans (1996). *In and about the World: Philosophical Studies of Science and Technology.* Albany, NY: State University of New York Press.
Radder, Hans (2003). "Technology and Theory in Experimental Science." In Hans Radder (ed.) *The Philosophy of Scientific Experimentation.* Pittsburgh: University of Pittsburgh Press, pp. 152 – 173.
Radder, Hans (2006). *The World Observed/The World Conceived.* Pittsburgh: University of Pittsburgh Press.
Radder, Hans (2009). "Science, Technology and the Science-Technology Relationship." In A. Meijers (ed.) *Philosophy of Technology and Engineering Sciences.* Amsterdam: Elsevier, pp. 65 – 91. DOI: 10.1016/B978-0-444-51667-1.50007 – 0.
Radder, Hans (2012). *The Material Realization of Science: From Habermas to Experimentation and Referential Realism.* Boston: Springer. DOI: 10.1007/978-94-007-4107-2.
Raven, Michael (2011). "There Is a Problem of Change." *Philosophical Studies*, 149: 77 – 96.
Raven, Michael (2016). "Fundamentality without Foundations." *Philosophy and Phenomenological Research*, 93(3): 607 – 626. DOI: 10.1111/phpr.12200.
Rehg, William (2009a). *Cogent Science in Context.* Cambridge: MIT Press.
Rehg, William (2009b). "Cogency in Motion: Critical Contextualism and Relevance." *Argumentation*, 23: 39 – 59.
Rehg, William (2009c). "Crossing Boundaries: Contexts of Practice as Common Goods." Conference paper for *Society for the Philosophy of Scientific Practices*, Minneapolis, presented June 18 – 20, 2009.
Reichenbach, Hans (1949). *The Theory of Probability.* Berkeley and Los Angeles: University of California Press.
Reichenbach, Hans (1956). *The Direction of Time.* Berkeley, CA: University of California Press.
Rescher, Nicholas (1967). *The Philosophy of Leibniz.* Upper Saddle River: Prentice-Hall.
Rescher, Nicholas (1996). *Process Metaphysics: An Introduction to Process Philosophy.* New York: Suny Press.
Rescher, Nicholas (2000). *Process Philosophy: A Survey of Basic Issues.* Pittsburgh: University of Pittsburgh Press.
Riffert, Franz G. (2008). "Whitehead's Process Philosophy as Scientific Metaphysics." In T. Eastman and H. Keeton (eds.) *Physics and Whitehead: Quantum Process and Experience.* Albany, NY: State University of New York Press, pp. 199 – 222.
Robinson, Denis (1982). "Re-identifying Matter." *Philosophical Review*, 81: 317 – 342.

Robinson, Howard (1974). "Prime Matter in Aristotle." *Phronesis*, 19: 168–188.

Robinson, Howard (2018). "Substance." *The Stanford Encyclopedia of Philosophy* (Spring 2020 Edition). Edward N. Zalta (ed.). URL: https://plato.stanford.edu/archives/spr2020/entries/substance/ (last accessed 18 January 2023).

Rosen, Joe (2008). "The Primacy of Asymmetry over Symmetry in Physics." In T. Eastman and H. Keeton (eds.) *Physics and Whitehead: Quantum Process and Experience*. Albany, NY: State University of New York Press, pp. 129–135.

Ross, Lauren (forthcoming). "Explanation in Contexts of Causal Complexity: Lessons from Psychiatric Genetics." *Minnesota Studies in Philosophy of Science* volume entitled *From Biological Practice to Scientific Metaphysics*.

Rychter, Pablo (2009). "There Is No Puzzle about Change." *Dialectica*, 63(1): 7–22.

Saatsi, Juha T. (2005). "On the Pessimistic Induction and Two Fallacies." *Philosophy of Science*, 72(5): 1088–1098. DOI: 10.1086/508959.

Saatsi, Juha T. (2015). "Replacing Recipe Realism." *Synthese*, 194(9): 3233–3244.

Saatsi, Juha T. (2016). "Dynamical Systems Theory and Explanatory Indispensability." *Philosophy of Science*, 84(5): 892–904.

Salmon, Nathan (1979). "How Not to Derive Essentialism from the Theory of Reference." *Journal of Philosophy*, 76: 703–725.

Salmon, Nathan (1981). *Reference and Essence*. Princeton, NJ: Princeton University Press.

Salmon, Nathan (2003). "Naming, Necessity, and Beyond." *Mind*, 112: 475–492. Reprinted in Salmon 2005, 377–397.

Salmon, Nathan (2005). *Reference and Essence*, 2nd edition with added appendices. Amherst, NY: Prometheus Books.

Salmon, Wesley C. (1970). "Statistical Explanation." In R. Colodny (ed.) *The Nature and Function of Scientific Theories*. Pittsburgh: University of Pittsburgh Press, pp. 173–231.

Salmon, Wesley C. (1977). "Laws, Modalities and Counterfactuals." *Synthese*, 35: 191–229.

Salmon, Wesley C. (1978). "Why Ask "Why?"? An Inquiry Concerning Scientific Explanation." *Proceedings and Addresses of the American Philosophical Association*, 51(6): 683–705.

Salmon, Wesley C. (1979a). *Hans Reichenbach: Logical Empiricist*. Dordrecht: Reidel.

Salmon, Wesley C. (1979b). "The Philosophy of Hans Reichenbach." In Salmon 1979a, 1–84. DOI: 10.1007/978-94-009-9404-1_1.

Salmon, Wesley C. (with Merrilee H. Salmon) (1979c). "Alternative Models of Scientific Explanation." *American Anthropologist*, 81(1): 61–74. DOI: 10.1525/aa.1979.81.1.02a00050.

Salmon, Wesley C. (1984). *Scientific Explanation and the Causal Structure of the World*. Princeton, NJ: Princeton University Press.

Salmon, Wesley C. (1990). "Scientific Explanation: Causation and Unification." *Critica*, 22(66): 3–21. Reprinted in Salmon 1998, 68–78.

Salmon, Wesley C. (1994). "Carnap, Hempel, and Reichenbach on Scientific Realism." In W. Salmon and G. Wolters (eds.) *Logic, Language, and the Structure of Scientific Theories*. Pittsburgh and Konstanz: University of Pittsburgh Press and Universitätsverlag Konstanz, pp. 237–254. Reprinted in Salmon 2005, pp. 19–30.

Salmon, Wesley C. (1994). "Causality without Counterfactuals." *Philosophy of Science*, 61(2): 297–312.

Salmon, Wesley C. (1998). *Causality and Explanation*. Oxford: Oxford University Press.

Salmon, Wesley C. (2005). *Reality and Rationality*. Phil Dowe and Merrilee H. Salmon (eds.). New York: Oxford University Press.

Sargent, C.L. (1981). "The Repeatability of Significance and the Significance of Repeatability." *European Journal of Parapsychology*, 3: 423–433.
Schaffer, Jonathan (2003). "Is There a Fundamental Level?" *Noûs*, 37(3): 498–517. DOI: 10.1111/1468-0068.00448.
Schaffer, Jonathan (2004). "Two Conceptions of Sparse Properties." *Pacific Philosophical Quarterly*, 85(1): 92–102. DOI: 10.1111/j.1468-0114.2004.00189.x.
Schaffer, Jonathan (2009). "On What Grounds What." In David Manley, David Chalmers, and Ryan Wasserman (eds.), *Metametaphysics: New Essays on the Foundations of Ontology*. Oxford: Oxford University Press, pp. 347–383.
Schaffer, Jonathan (2010a). "Monism: The Priority of the Whole." *Philosophical Review*, 119(1): 31–76. DOI: 10.1215/00318108-2009-025.
Schaffer, Jonathan (2010b). "The Least Discerning and Most Promiscuous Truthmaker." *Philosophical Quarterly*, 60(239): 307–324. DOI: 10.1111/j.1467-9213.2009.612.x.
Schechtman, Marya (1996). *The Constitution of Selves*. Ithaca: Cornell University Press.
Schechtman, Marya (2001). "Empathic Access: The Missing Ingredient in Personal Identity." *Philosophical Explorations*, 4(2): 94–110.
Schmidt, Stefan (2009). "Shall We Really Do It Again? The Powerful Concept of Replication Is Neglected in the Social Sciences." *Review of General Psychology*, 13(2): 90–100. DOI: 10.1037/a0015108.
Schroer, Jeanine W. and Robert Schroer (2014). "Getting the Story Right: A Reductionist Narrative Account of Personal Identity." *Philosophical Studies*, 171: 445–469.
Schupbach, Jonah N. (2015). "Robustness, Diversity of Evidence, and Probabilistic Independence." In U. Mäki, S. Ruphy, G. Schurz, and I. Votsis (eds.) *Recent Developments in the Philosophy of Science: EPSA13*. Helsinki: Springer, pp. 305–316.
Schupbach, Jonah N. (2016). "Robustness Analysis as Explanatory Reasoning." *British Journal for the Philosophy of Science*, 69(1): 275-300.
Schwinger, Julian (1951). "The Theory of Quantized Fields I." *Physical Review* 82(6): 914–927.
Seibt, Johanna (1990). *Towards Process Ontology: A Critical Study of Substance-Ontological Premises*, Ph.D. Thesis, University of Pittsburgh; UMI Microfiche Publication.
Seibt, Johanna (1996a). "Non-Countable Individuals: Why One and the Same Is Not One and the Same." *Southwest Philosophy Review*, 12: 225–237.
Seibt, Johanna (1996b). "The Myth of Substance and the Fallacy of Misplaced Concreteness." *Acta Analytica*, 15: 61–76.
Seibt, Johanna (1996c). "Existence in Time: From Substance to Process." In J. Faye (ed.) *Perspectives on Time. Boston Studies in Philosophy of Science*. Dordrecht: Kluwer, pp. 143–182.
Seibt, Johanna (2004a). "Free Process Theory: Towards a Typology of Processes." *Axiomathes*, 14: 23–57.
Seibt, Johanna (2004b). *General Process Theory*. Habilitationsschrift at the University of Konstanz.
Seibt, Johanna (2007). "Beyond Endurance and Perdurance: Recurrent Dynamics." In C. Kanzian (ed.) *Persistence*. Frankfurt: Ontos, pp. 133–165.
Seibt, Johanna (2009). "Forms of Emergence in General Process Theory." *Synthese*, 166: 479–517.
Seibt, Johanna (2010). "Particulars." In R. Poli and J. Seibt (eds.) *Theory and Applications of Ontology: Philosophical Perspectives*. Heidelberg/New York: Springer, pp. 23–56.
Seibt, Johanna (2015). "Non-Transitive Parthood, Leveled Mereology, and the Representation of Emergent Parts of Processes." *Grazer Philosophische Studien*, 91: 165–190.

Seibt, Johanna (2017). "Process Philosophy." *The Stanford Encyclopedia of Philosophy* (Summer 2020 Edition). Edward N. Zalta (ed.). URL: https://plato.stanford.edu/archives/sum2020/entries/process-philosophy/ (last accessed 18 January 2023).

Seibt, Johanna (2018). "What Is a Process? Modes of Occurrence and Forms of Dynamicity in *General Process Theory*." In R. Stout (ed.) *Processes, Experiences, and Actions*. Oxford: Oxford University Press.

Sellars, Wilfrid (1952). "Particulars." *Philosophy and Phenomenological Research*, 13: 184–199.

Sellars, Wilfrid (1981). "Foundations for a Metaphysics of Pure Process." *The Monist*, 64(1): 3–90.

Shoemaker, Sarah (2008). "Persons, Animals, and Identity." *Synthese*, 163: 313–324.

Shoemaker, Sarah (2011). "On What We Are." In S. Gallagher (ed.) *The Oxford Handbook of the Self*. Oxford: Oxford University Press, pp. 352–371.

Shumener, Erica (2019a). "Building and Surveying: Relative Fundamentality in Karen Bennett's Making Things Up." *Analysis*, 79(2): 303–314.

Shumener, Erica (2019b). "Laws of Nature, Explanation, and Semantic Circularity." *British Journal for the Philosophy of Science*, 70(3): 787–815.

Shumener, Erica (2020a). "Explaining Identity and Distinctness." *Philosophical Studies*, 177(7): 2073–2096.

Shumener, Erica (2020b). "Identity." In M.J. Raven (ed.) *Routledge Handbook of Metaphysical Grounding*. London: Routledge, pp. 413–424.

Shumener, Erica (2021). "Humeans Are Out of This World." *Synthese*, 198(6): 5897–5916.

Sider, Theodore (1996). "All the World's a Stage." *Australasian Journal of Philosophy*, 74: 433–453.

Sider, Theodore (2000). "The Stage View and Temporary Intrinsics." *Analysis*, 60: 84–88.

Sider, Theodore (2001). *Four-Dimensionalism*. Oxford: Oxford University Press.

Sider, Theodore (2011). *Writing the Book of the World*. Oxford University Press.

Silagade, Zurab (2005). "Zeno Meets Modern Science." *Acta Physica Polonica B*, 36(10): 2887–2930.

Simons, Peter (1987). *Parts: A Study in Ontology*. Oxford: Clarendon Press.

Simons, Peter (2000). "Identity through Time and Trope Bundles." *Topoi*, 19: 147–155.

Simons, Peter and Joseph Melia (2000). "Continuants and Occurrents." *Proceedings of the Aristotelian Society, Supplementary Volumes* 74: 59–75, 77–92.

Smart, John J.C. (1963). *Philosophy and Scientific Realism*. London: Routledge & Kegan Paul.

Sober, Elliott (2015). "Two Cornell Realisms: Moral and Scientific." *Philosophical Studies*, 172(4): 905–924. DOI: 10.1007/s11098-014-0300-5.

Stahl, Georg Ernst (1744). *Materia Medica: das ist: Zubereitung, Krafft, und Wurckung, derer sonderlich durch chymische Kunst erfundenen Artzneyen*. Vol 1, 2.

Stanford, P. Kyle (2001). "Refusing the Devil's Bargain: What Kind of Underdetermination Should We Take Seriously?" *Philosophy of Science*, 68: S1–S12.

Stanford, P. Kyle (2006). *Exceeding Our Grasp: Science, History, and the Problem of Unconceived Alternatives*. New York: Oxford University Press.

Stapp, Henry P. (2008). "Whiteheadian Process and Quantum Theory." In T. Eastman and H. Keeton (eds.) *Physics and Whitehead: Quantum Process and Experience*. Albany, NY: State University of New York Press, pp. 92–102.

Steinle, Friedrich (2016). "Stability and Replication of Experimental Results: A Historical Perspective." In Harald Atmanspacher and Sabine Maasen (eds.) *Reproducibility: Principles, Problems, Practices, and Prospects*. Hoboken, NJ: Wiley and Sons Inc., pp. 39–68. DOI: 10.1002/9781118865064.ch3.

Strawson, Galen (2008). "Against Narrativity." In Galen Strawson, *Real Materialism and Other Essays*. Oxford: Oxford University Press, pp.189–207.

Strawson, Galen (2011). *Locke on Personal Identity: Consciousness and Concernment.* Princeton, NJ: Princeton University Press.
Strawson, Peter F. (1959). *Individuals: An Essay in Descriptive Metaphysics.* London: Routledge.
Strawson, Peter F. (1966a). "XII.—Particular and General." *Proceedings of the Aristotelian Society,* 54(1): 233–260.
Strawson, Peter F. (1966b). "The Stratification of Behaviour: A System of Definitions Propounded and Defended." *Philosophical Quarterly,* 16(65): 389.
Stuewer, Roger H. (1994). "The Origin of the Liquid-Drop Model and the Interpretation of Nuclear Fission." *Perspectives on Science,* 2: 76–129.
Suárez, Mauricio (2003). "Scientific Representation: Against Similarity and Isomorphism." *International Studies in the Philosophy of Science,* 17(3): 225–244.
Suárez, Mauricio (2013). "Interventions and Causality in Quantum Mechanics." *Erkenntnis,* 78(2): 199–213.
Suárez, Mauricio (2010). "Scientific Representation." *Philosophy Compass* 5(1): 91–101.
Suppe, Frederick (1989). *The Semantic View of Theories and Scientific Realism.* Urbana and Chicago: University of Illinois Press.
Swinburne, Richard (1984). "Personal Identity: The Dualist Theory." In S. Shoemaker and R. Swinburne, *Personal Identity.* Oxford: Blackwell, pp. 1–66.
Tahko, Tuomas E. (2013). "Truth-grounding and Transitivity." *Thought: A Journal of Philosophy,* 2(4): 332–340. DOI: 10.1002/tht3.94.
Tahko, Tuomas E. (2014). "Boring Infinite Descent." *Metaphilosophy,* 45(2): 257–269. DOI: 10.1111/meta.12084.
Tanaka, Yutaka (2008). "The Individuality of a Quantum Event: Whitehead's Epochal Theory of Time and Bohr's Framework of Complementarity." In T. Eastman and H. Keeton (eds.) *Physics and Whitehead: Quantum Process and Experience.* Albany, NY: State University of New York Press, pp. 164–179.
Teller, Paul (2004). "How We Dapple the World." *Philosophy of Science,* 71(4): 425–447.
Tye, Michael (1975). "The Adverbial Theory: A Defense of Sellars against Jackson." *Metaphilosophy,* 6: 136–143.
Unger, Peter (1979). "I Do Not Exist." In G.F. MacDonald (ed.) *Perception and Identity.* London: Macmillan, pp. 235–251.
Unger, Peter (1990). *Identity, Consciousness, and Value.* Oxford: Oxford University Press.
Van Fraassen, Bas (1980). *The Scientific Image.* Oxford: Oxford University Press.
Van Fraassen, Bas (1989). *Laws and Symmetry.* Oxford University Press.
Van Fraassen, Bas (1995). "Interpretation of Science, Science as Interpretation." In J. Hilgevoord (ed.) *Physics and Our View of the World.* Cambridge: Cambridge University Press, pp. 188–225.
Van Fraassen, Bas (2006). "Structure: Its Shadow and Substance." *British Journal for the Philosophy of Science,* 57: 275–307.
Van Fraassen, Bas (2007). "Structuralism(s) About Science: Some Common Problems." *Proceedings of the Aristotelian Society,* 81: 45–61.
Van Fraassen, Bas (2008a). *Scientific Representation: Paradoxes of Perspective.* Oxford: Oxford University Press.
Van Fraassen, Bas (2008b). *Scientific Representation.* Oxford: Oxford University Press.
Van Fraassen, Bas (2014). "One or Two Gentle Remarks about Hans Halvorson's Critique of the Semantic View." *Philosophy of Science,* 81: 276–283.
Van Inwagen, Peter (1990). *Material Beings.* Ithaca: Cornell University Press.

Van Inwagen, Peter (1997). "Materialism and the Psychological-Continuity Account of Personal Identity." *Philosophical Perspectives*, 11: 305–319.
Vendler, Zeno (1957). "Verbs and Times." *Philosophical Review*, 66(2): 143–160.
Vinci, Thomas (1981). "Sellars and the Adverbial Theory of Sensation." *Canadian Journal of Philosophy*, 11: 199–217.
Votsis, Ioannis (2003). "Is Structure Not Enough?" *Philosophy of Science*, 70: 879–890.
Votsis, Ioannis (2005). "The Upward Path to Structural Realism." *Philosophy of Science*, 72: 1361–1372.
Wasserman, Ryan (2003). "The Argument from Temporary Intrinsics." *Australasian Journal of Philosophy*, 81: 413–419.
Wasserman, Ryan (2004). "Framing the Debate over Persistence." *Metaphysica*, 5: 67–80.
Wasserman, Ryan (2005). "Humean Supervenience and Personal Identity." *Philosophical Quarterly*, 55: 582–593.
Wasserman, Ryan (2006). "The Problem of Change." *Philosophy Compass*, 1: 48–57.
Wasserman, Ryan (2016). "Theories of Persistence." *Philosophical Studies*, 173: 243–250.
Watsuji, Tetsurō (1988 [1935]). *Fūdo*. Translated in Geoffrey Bownas (trans.), *Climate and Culture: A Philosophical Study*. New York: Greenwood Press.
Weisberg, Michael and Kenneth Reisman (2008). "The Robust Volterra Principle." *Philosophy of Science*, 75(1): 106–131.
Weizsäcker, Carl Friedrich von (1935). "Zur Theorie der Kernmassen." *Zeitschrift für Physik*, 96: 431–458.
Wiggins, David (1980). *Sameness and Substance*, Oxford: Blackwell.
Wiggins, David (2001). *Sameness and Substance Renewed*. Cambridge: Cambridge University Press.
Wiggins, David (2016a). *Essays on Identity and Substance*. Oxford: Oxford University Press.
Wiggins, David (2016b). "Activity, Process, Continuant, Substance, Organism." *Philosophy*, 91(2): 269–280.
Wiggins, David (2016c). *Continuants: Their Activity, Their Being, and Their Identity*. Oxford: Oxford University Press.
Williams, Bernard (1956–57). "Personal Identity and Individuation." *Proceedings of the Aristotelian Society* 57: 229–252. Reprinted in Bernard Williams, *Problems of the Self*. Cambridge: Cambridge University Press 1973, pp. 229–252.
Williams, Bernard (1970). "The Self and the Future." *Philosophical Review*, 79(2): 161–180.
Wilson, Jessica M. (2012). "Fundamental Determinables." *Philosophers' Imprint*, 12(4): 1–17.
Wilson, Jessica M. (2014). "No Work for a Theory of Grounding." *Inquiry*, 57(5–6): 535–579. DOI: 10.1080/0020174X.2014.907542.
Wilson, Jessica M. (2016). "The Unity and Priority Arguments for Grounding." In Ken Aizawa and Carl Gillett (eds.) *Scientific Composition and Metaphysical Ground*. Basinstoke: Palgrave MacMillan, pp.171–204.
Wolff, Johanna (2012). "Do Objects Depend on Structures?" *British Journal for the Philosophy of Science*, 63(3): 607–625. DOI: 10.1093/bjps/axr041.
Woodward, James (2015). "Normative Theory and Descriptive Psychology in Understanding Causal Reasoning: The Role of Interventions and Invariance." In W. Gonzalez (ed.) *Philosophy of Psychology: Causality and Psychological Subject*. Berlin: De Gruyter, pp. 71-104.
Woodward, James (2019). "Causal Attribution, Counterfactuals and Disease Interventions." Preprint available at Philsci Archive, philsci-archive.pitt.edu/16037/ (last accessed June 1, 2020).
Woodward, Jim (1999). "Causal Interpretation in Systems of Equations." *Synthese*, 121(1–2): 199–247.

Woodward, Jim (2003). *Making Things Happen: A Theory of Causal Explanation.* Oxford: Oxford University Press.
Woodward, Jim (2014a). "From Handles to Interventions: Commentary on R.G. Collingwood, 'The So-Called Idea of Causation'." *International Journal of Epidemiology,* 43(6):1–6.
Woodward, Jim (2014b). "A Functional Theory of Causation." *Philosophy of Science* 81: 691–713.
Worrall, John (1989). "Structural Realism: The Best of Both Worlds?" *Dialectica,* 43: 99–124. Reprinted in D. Papineau (ed.) *The Philosophy of Science.* Oxford: Oxford University Press 1996, pp. 139–165.
Worrall, John (2007). "Miracles and Models: Why Reports of the Death of Structural Realism May Be Exaggerated." *Royal Institute of Philosophy Supplements,* 82(61): 125–154.
Wray, K. Brad (2007). "A Selectionist Explanation of the Success and Failures of Science." *Erkenntnis,* 67(1): 81–89. DOI: 10.1007/s10670-007-9046-1.
Wray, K. Brad (2010). "Selection and Predictive Success." *Erkenntnis,* 72(3): 365–377. DOI: 10.1007/s10670-009-9206-6.
Yuasa, Yasuo (1987). *The Body: Toward an Eastern Mind-Body Theory.* Ed. T.P. Kasulis. Trans. Nagatomo Shigenori and T.P. Kasulis. New York: State University of New York Press.
Yukawa, H. (1935). "On the Interaction of Elementary Particles, I." *Proceedings of the Physico-Mathematical Society of Japan,* 17: 48–57.
Zalta, Edward (2006). "Essence and Modality." *Mind,* 115: 659–693.
Zemach, Eddie (1970). "Four Ontologies." *The Journal of Philosophy,* 23(8): 231–247.
Zimmerman, Dean (1996). "Could Extended Objects Be Made Out of Simple Parts? An Argument for 'Atomless Gunk'." *Philosophy and Phenomenological Research,* 56(1): 1–29. DOI: 10.2307/2108463.
Zimmerman, Dean (1998). "Temporary Intrinsics and Presentism." In P. van Inwagen and D. Zimmerman (eds.) *Metaphysics, the Big Questions.* Oxford: Blackwell, pp. 206–219.

Index

Aristotle 1, 34 f., 40 f., 66, 73 f., 80, 96, 149, 166
Atomism 56, 75

Becher, Johannes 75, 95, 97 f., 114
Bohr, Niels 32, 157, 160 f., 165
Brownian motion 2, 56, 126-134, 141, 144

Candle flame 1, 38 f., 45, 47-51, 54, 83 f., 93-124, 126, 131 f., 171
Carnap, Rudolf 15 f., 59
Continuant 8, 30, 96, 97, 109, 149 f., 192
Continuity 3 f., 21-25, 28-30, 32
Continuity argument 4 f., 9, 17 f., 21, 25, 28, 31 f., 48, 86
Countable 8 f., 20 f., 96 f., 105 f., 109, 114, 127 f., 141 f.

Dirac, Paul 146
Dynamic shape 10, 44, 66, 80, 97, 106, 112, 122 f.

Empirical adequacy 17

Faraday, Michael 98 f., 101-104, 107-110, 114, 117, 120
Feyerabend, Paul 12, 15 f.
Finkelstein, David 4, 7, 11, 31, 160

Heraclitus 6, 74, 82, 95

Kant, Immanuel 14

Measurable 8, 13-15, 20 f., 30, 66, 76, 82, 93, 106, 112, 123, 127, 138, 141

Nucleus 31, 62, 66-69, 98, 147-172

Occurrent 3, 8, 15, 19, 30, 79, 81, 96, 106, 109, 112, 126

Parmenides 6, 74, 79
Perrin, Jean Baptiste 2, 67, 69, 124, 126-130, 132-145
Persistence 37-45, 47, 50 f., 55, 95 f., 102 f., 158
Persistent 34, 39, 42-45, 47, 50-55, 57 f., 90, 95, 101, 158
Phlogiston 1, 75, 94-99
Plenum 56, 75, 144

Reference 16, 22, 41 f., 59-69, 72 f., 79, 83, 87, 102, 121 f., 124, 126, 160, 164, 170
Robustness 2 f., 35, 37 f., 45, 47, 49-58, 68-70, 77, 80, 85, 124, 126-128, 132 f., 135, 150, 157, 164, 172, 174
Rutherford, Ernest 1, 26

Seibt, Johanna 3, 5, 7-11, 15, 35, 41 f., 59, 62-64, 66, 73, 123, 148 f.
Stability 1, 6, 14, 35-40, 43-55, 57-59, 64-72, 76, 79, 82-90, 99, 101 f., 104, 111 f., 119 f., 123, 128, 149, 151, 153-155, 159, 162 f., 167-169, 174
Stahl, Ernst 75, 95, 98, 114
Staticity 1, 6, 13 f., 31 f., 34-36, 38-44, 46-48, 50-55, 57-59, 61, 64, 67 f., 70-72, 79-82, 84-86, 88, 94, 99, 102, 104, 111, 122, 124, 126, 128, 144 f., 147-150, 160, 168 f., 172
Symmetry 72, 81 f., 101 f., 110

Unchanging room 5, 12, 17
Underlier 34-37, 41, 45, 48 f., 54 f., 57 f., 63, 67, 70, 73, 81, 85 f., 102, 115, 120, 123-125, 135, 144 f., 167, 169, 172
Underlier argument 1, 17, 34-40, 42, 44-50, 55, 57-59, 64 f., 68, 71 f., 80, 84-86, 126, 158, 166, 168
Unification 49, 68-71, 77, 80, 85, 120, 149

Zeno's paradoxes 79, 79 f, 80

www.ingramcontent.com/pod-product-compliance
Lightning Source LLC
Chambersburg PA
CBHW050526170426
43201CB00013B/2097